ADVANCES IN EXPERIMENTAL MEDICINE AND BIOLOGY

PURINE METABOLISM IN MAN
Enzymes and Metabolic Pathways

PURINE METABOLISM IN MAN

Enzymes and Metabolic Pathways

Edited by

Oded Sperling and Andre De Vries

Division of Metabolic Disease
Rogoff-Wellcome Medical Research Institute
Beilinson Hospital
Petah-Tikva, Israel

and

James B. Wyngaarden

Department of Medicine
Duke University Medical Center
Durham, North Carolina

PLENUM PRESS • NEW YORK-LONDON

Library of Congress Cataloging in Publication Data

International Symposium on Purine Metabolism in Man,
   Tel-Aviv, 1973.
   Purine metabolism in man: enzymes and metabolic pathways.

   (Advances in experimental medicine and biology, v. 41)
   Vol. 2 has subtitle: Biochemistry and pharmacology of uric acid metabolism.
   Includes bibliographical references.
   1. Purine metabolism—Congresses. 2. Uric acid metabolism—Congresses. I. Sperling, Oded, ed.
II. De Vries, André, 1911-   ed. III. Wyngaarden, James B., ed. IV. Title. V. Series. [DNLM:
1. Purine-pyrimidine metabolism, Inborn errors—Congresses. 2. Purines—Metabolism—
Congresses. W1AD559 v. 41 / QU58 I635p 1973]
QP801.P8I56   1973   612'.39   73-21823
ISBN 0-306-39095-7 (v. 1)

First half of the proceedings of the International Symposium on Purine Metabolism in Man,
held in Tel Aviv, Israel, June 17-22, 1973

© 1974 Plenum Press, New York
A Division of Plenum Publishing Corporation
227 West 17th Street, New York, N.Y. 10011

United Kingdom edition published by Plenum Press, London
A Division of Plenum Publishing Company, Ltd.
Davis House (4th Floor), 8 Scrubs Lane, Harlesden, London, NW10 6SE, England

PREFACE

Gout and uric acid lithiasis are known to have
affected mankind for thousands of years. It is only
recently, however, that great progress has been made
in the understanding of the processes involved in purine
metabolism and its disorders in man. The key enzymes
active in the various pathways of purine synthesis and
degradation have become known and their properties are
the subject of intensive study. Major contributions to
the knowledge of normal purine metabolism in man have
derived from the study of inborn errors in patients with
purine disorders, specifically complete and partial
hypoxanthine-guanine phosphoribosyltransferase deficiency.
Mutations of other enzymes involved in purine metabolism
are being discovered. A great step forward has been
made in the treatment of gout with the introduction of
uricosuric drugs and more recently of the hypoxanthine
analogue allopurinol, a synthetic xanthine oxidase
inhibitor. Furthermore, the complex nature of the
renal handling of uric acid excretion, although still
posing difficult problems, appears to approach clari-
fication.

In view of the intensive research on purine metabolism going on in various laboratories all over the
world it was felt by several investigators in this field
that the time was appropriate to convene a symposium
dedicated to this subject. An International Symposium
on Purine Metabolism in Man, the first of its kind, was
therefore organized and held in Tel-Aviv, June 17 to
22, 1973. The meeting dealt with the various aspects of
purine metabolism and its disorders - biochemistry,
enzyme mutations, genetics, methodology, clinical aspects
and treatment. This volume contains the full reports
of all communications made at the symposium. Its publi-

cation was made possible by the joint effort of the contributors and the organizing and advisory committees of the congress.

The special aid by the various sponsors, the Israel Ministry of Health, the Tel-Aviv University, the Israel Academy of Sciences and Humanities, the Wellcome Research Laboratories of the Burroughs Wellcome Co. USA, and the excellent organization by Kenes, the congress organizers, are gratefully acknowledged.

The Editors

CONTENTS OF VOLUME 41A

ENZYMES AND METABOLIC PATHWAYS

IN PURINE METABOLISM

Purine Phosphoribosyltransferases

Nucleoside and Nucleotide Metabolism

MUTATIONS AFFECTING PURINE METABOLISM

Properties of HGPRT and APRT in HGPRT

Deficient Blood Cells

Clinical Manifestations and Genetic Aspects

Purine Metabolism and Erythrocyte PRPP Content

in Heterozygotes for HGPRT Deficiency

Mutants of PRPP Synthetase

APRT Deficiency

Xanthinuria

Glycogen Storage Disease

CONTENTS OF VOLUME 41B

GOUT

Etiology of Purine Overproduction in Gout

Effects of Diet, Weight, and Stress
on Purine Metabolism

Relationship Between Carbohydrate,
Lipid and Purine Metabolism

Renal Disease and Uric Acid Lithiasis

in Urate Overproduction

Urate Deposition in Tissue

Adenine Therapy in the

Lesch-Nyhan Syndrome

METHODOLOGY

ABBREVIATIONS

AMP	Adenylic acid
APRT	Adenine phosphoribosyltransferase
BSP	Bromsulphalein
BBR	Benzbromarone
BZD	Benziodarone
CRM	Cross reactive material
2,3-DPG	2,3-Diphosphoglycerate
FGAR	Formylglycinamide ribonucleotide
GAR	Glycinamide ribonucleotide
GFR	Glomerular filtration rate
GMP	Guanylic acid
GPRT	Guanine phosphoribosyltransferase
GSD	Glycogen storage disease
HPRT	Hypoxanthine phosphoribosyltransferase
HGPRT	Hypoxanthine-guanine phosphoribosyltransferase
IMP	Inosinic acid
LNS	Lesch-Nyhan syndrome
NAD	Nicotinamide adenine dinucleotide
ODC	Orotidylate decarboxylase
OPRT	Orotate phosphoribosyltransferase
Pb	Probenecid
PHA	Phytohaemagglutinine
PRA	Phosphoribosylamine
PRPP	5-Phosphoribosyl-1-pyrophosphate
PRT	Phosphoribosyltransferase
PZA	Pyrazinamide
R5P	Ribose-5-phosphate
UA	Uric acid
XDH	Xanthine dehydrogenase
XMP	Xanthylic acid

ENZYMES AND
METABOLIC PATHWAYS
IN PURINE METABOLISM

Purine
Phosphoribosyltransferases

HUMAN HYPOXANTHINE-GUANINE PHOSPHORIBOSYLTRANSFERASE (HGPRT):

PURIFICATION AND PROPERTIES

W. J. Arnold, R. V. Lamb III, and W. N. Kelley

Department of Medicine, Duke University Medical Center

Durham, North Carolina   27710

Hypoxanthine-guanine phosphoribosyltransferase (HGPRT) (E.C. 2.4.2.8.) catalyzes the formation of guanosine-5'-monophosphate and inosine-5'-monophosphate from 5'-phosphoribosyl-1-pyrophosphate (PP-ribose-P) and the purine bases guanine and hypoxanthine, respectively. Although originally assigned simply a "salvage" function in purine metabolism, the discovery of a virtually complete, X-linked deficiency of HGPRT associated with hyperuricemia, hyperuricaciduria and a bizarre neurologic syndrome (Lesch-Nyhan syndrome) has led to a reevaluation of the importance of HGPRT in the regulation of purine metabolism and central nervous system function (Seegmiller, Rosenbloom and Kelley, 1967). Therefore, an analysis of the structure of the normal enzyme is necessary to provide a basis for understanding the genetic lesion(s) producing altered or absent enzyme function. We have purified HGPRT to homogeneity from non-pooled, human male erythrocytes and have attempted to elucidate the nature of the electrophoretic heterogeneity of HGPRT activity observed during the purification (Arnold and Kelley, 1971; Kelley and Arnold, 1973).

Table 1 summarizes the results of the purification of HGPRT. Enzyme instability was a major problem and required the use of magnesium PP-ribose-P, dithiothreitol, sucrose, ampholytes, dimethylsulfoxide and bovine serum albumin at various times during the purification. Ion exchange chromatography on DEAE-cellulose resulted in a single peak of HGPRT activity which was pooled, concentrated by Amicon ultrafiltration (UM-10 membrane) and subjected to heat treatment at 60° in the presence of 5 mM $MgCl_2$ and 1 mM PP-ribose-P. Following this, low ionic strength dialysis against 2 mM Tris buffer pH 7.4 was done to reduce the ionic content of the solution. The sample was then immediately applied to an LKB Uniphor 7900 preparative isoelectric focusing column in which a 220 ml 0-40%

5

TABLE 1

PURIFICATION OF HYPOXANTHINE-GUANINE PHOSPHORIBOSYLTRANSFERASE FROM HUMAN ERYTHROCYTES (MALE)

Step	Specific Activity	Total Protein	Recovery	Purification
	nmoles/mg protein/hr	mg	% initial activity	-fold
1. Hemolysate	81	72,000		
2. DEAE eluate	9,470	399	64	177
3. Heat-treatment	29,206	85	42	361
4. Dialysis	36,245	67	42	447
5. Isofocusing				
I			4.2 ⎫	
II			5.8 ⎬ 11.2	
III			1.2 ⎭	
6. Sephadex G-100				
I	229,207	0.56	3.7 ⎫	8487
II	223,883	1.44	5.3 ⎬ 10.0	8022
III	83,000	0.47	1.0 ⎭	3141

(From Arnold and Kelley, 1971)

linear sucrose gradient, 1% in carrier ampholytes, pH 4 to 6, was poured. After 72 hours of isoelectric focusing at 5° the column was eluted from below and the pH and HGPRT activity of each 1 ml fraction was determined (Figure 1). Each peak of HGPRT activity (I, II, III) was pooled separately, concentrated by Amicon ultra-filtration and passed through a Sephadex G-100 column equilibrated in 50 mM Tris buffer pH 7.4, 0.5 mM dithiothreitol and 5% DMSO. Each electrophoretic variant and an impure preparation of HGPRT (Step 1) was found to have a Stokes radius of 36 Å and molecular weight of 68,000 by molecular sieve chromatography on Sephadex G-100 (Table 2). Analytical polyacrylamide electrophoresis of each electrophoretic variant at three different cross-linkages (5%, 10%, 15%) revealed only a single major protein band present in all three samples. The migration of each electrophoretic variant in polyacrylamide gels with 10% cross-linkage is shown in Figure 2. HGPRT activity could be assayed from each and correspon-ded to the position of the major protein band. Sodium dodecyl sulfate polyacrylamide gel electrophoresis with or without 1% β-mercaptoethanol demonstrated a single major band for each electro-phoretic variant with a molecular weight of approximately 34,000 and suggests that the subunits are associated by non-covalent forces

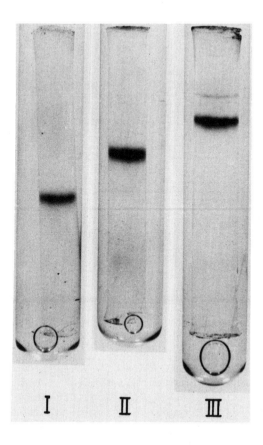

Fig. 2. Polyacrylamide gel electrophoresis of electrophoretic variants I, II and III.

acid composition of variants II and III show a striking similarity (Table 4).

The nature of the post-translational alteration responsible for the electrophoretic heterogeneity of HGPRT remains unclear. Differential sialation appears unlikely since exposure of the electrophoretic variants to neuraminidase does not liberate sialic acid nor alter their electrophoretic mobility. Also, the

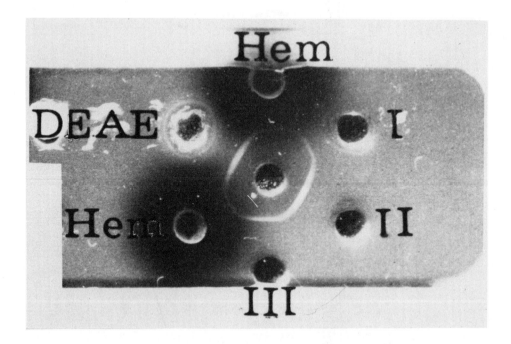

Fig. 3.   Immunodiffusion of normal human HGPRT. Peripheral wells
contain: HEM - hemolysate; I, II, III - electrophoretic variants
of HGPRT; DEAE - STEP 1 partially-purified HGPRT: Center well
contains mono-specific anti-HGPRT. Only a single line is present
for each preparation of HGPRT and all show reactions of identity.

association of the subunits into aggregates with different
molecular weights has been excluded by the identical elution
patterns on Sephadex G-100. Finally, we have been unable to detect
more than one peak of HGPRT activity in preparations of enzyme
prior to isoelectric focusing when either polyacrylamide gel or
agarose film electrophoresis is used (Figure 5). This observation
in conjunction with the interconvertability of the electrophoretic
variants shown in Figure 4 and the demonstrated absence of covalent
subunit linkage suggests an equilibrium between the subunits which
favors electrophoretic variant II.

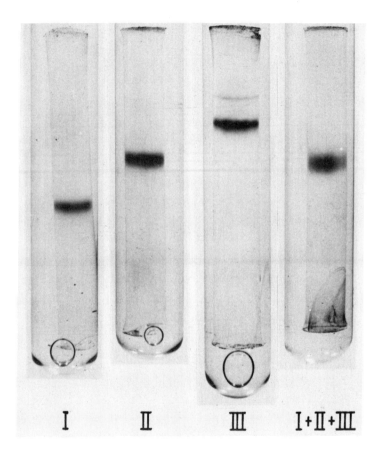

Fig. 4. Analytical polyacrylamide gel electrophoresis of HGPRT electrophoretic variants (I, II, III) and a mixture of the three (I+II+III).

TABLE 3

COMPARISON OF ISOELECTRIC POINTS AND RECOVERY OF PURIFIED
ELECTROPHORETIC VARIANTS OF HUMAN HGPRT

Electrophoretic Variant	Per Cent of Total HGPRT activity recovered[+]		Isoelectric Point[+]
	(mean ± S.D.)	Range	(mean ± S.D.)
I	31±11	15-41	5.65±0.06
II	45±4.5	38-52	5.8±0.06
III	25±12	11-39	6.01±0.09

[+] Data calculated from 7 separate but identical purifications
of HGPRT. (From Arnold and Kelley, 1973)

TABLE 4

HYPOXANTHINE-GUANINE PHOSPHORIBOSYLTRANSFERASE: AMINO ACID
COMPOSITION OF ELECTROPHORETIC VARIANTS II AND III

Amino Acid	Number of Amino Acid Residues per Subunit	
	II	III
Lysine	11	13
Histidine	8	8
Arginine	18	15
Aspartic Acid	27	24
Threonine	11	14
Serine	13	14
Glutamic Acid	30	29
Proline	18	19
Glycine	26	28
Alanine	22	21
Valine	20	18
Methionine	8	10
Isoleucine	8	8
Leucine	24	24
Tyrosine	7	8
Phenylalanine	12	12

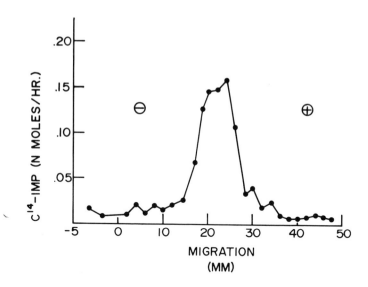

Fig. 5. Agarose film electrophoresis of HGPRT. Dialyzed normal human hemolysate was diluted 1:10 in 0.05 M Tris-HCl pH 8.45. An aliquot of 4 ul was applied to an agarose film strip equilibrated in the same buffer. Electrophoresis was done at 4° for 2 hours with 50 volts direct current. The agarose film was then cut into 2 mm strips (width-wise) which were then placed into the usual assay mixture and incubated for 2 hours at 37°.

REFERENCES

Arnold, W.J. and Kelley, W.N. 1971. Human hypoxanthine-guanine phosphoribosyltransferase: Purification and subunit structure. J. Biol. Chem. 246: 7398-7404.

Bakay, B. and Nyhan, W.L. 1971. The separation of adenine and hypoxanthine phosphoribosyltransferases isoenzymes by disc gel electrophoresis. Biochem. Genet. 5: 81-90.

Gutensohn, W. and Guroff, G. 1972. Hypoxanthine-guanine phosphoribosyltransferase from rat brain: Purification, kinetic properties, development and distribution. J. Neurochem. 19: 2139-2150.

Kelley, W.N. and Arnold, W.J. Human hypoxanthine-guanine phosphoribosyltransferase: Studies on the normal and mutant forms of the enzyme. Fed. Proc. (in press).

Muller, M.M. and Debrovits, H. 1972. A simple and rapid method
    for the separation and characterization of hypoxanthine-guanine
    phosphoribosyltransferase isoenzymes. Prep. Biochem. $\underline{2}$: 375-
    389.

Rosenbloom, F.M., Kelley, W.N., Miller, J.M., Henderson, J.F. and
    Seegmiller, J.E. 1967. Inherited disorder of purine metabol-
    ism: Correlation between central nervous system dysfunction
    and biochemical defects. J.A.M.A. $\underline{202}$: 175-177.

Rubin, C.S., Dancis, J., Yip, L.C., Nowinski, R.C. and Balis, M.E.
    1971. Purification of IMP: Pyrophosphate phosphoribosyltrans-
    ferases, catalytically incompetent enzymes in Lesch-Nyhan
    disease. Proc. Nat. Acad. Sci. $\underline{68}$: 1461-1464.

STABILIZATION BY PRPP OF CELLULAR PURINE PHOSPHORIBOSYL-TRANSFERASES AGAINST INACTIVATION BY FREEZING AND THAWING. STUDY OF NORMAL AND HYPOXANTHINE-GUANINE PHOSPHORIBOSYL-TRANSFERASE DEFICIENT HUMAN FIBROBLASTS

E. Zoref, O. Sperling and A. de Vries

Tel-Aviv University Medical School, Department
of Pathological Chemistry, Sheba Medical Center,
Tel-Hashomer and Rogoff-Wellcome Medical
Research Institute, Beilinson Hospital,
Petah Tikva, Israel

Hypoxanthine-guanine phosphoribosyltransferase
(HGPRT, EC 2.4.2.8) and adenine phosphoribosyltransferase
(APRT, EC 2.4.2.7) catalyze the salvage pathway formation
of purine nucleotides from the corresponding preformed
purine bases by reacting them with a common substrate,
5-phosphoribosyl-1-pyrophosphate (PRPP) (1,2). A deficiency of HGPRT in man causes excessive de novo production
of purines. When the enzyme deficiency is virtually
complete, it is associated with the Lesch-Nyhan syndrome
(LNS) (3,4), when partial, it is associated with severe
gout (5).

Freezing and thawing or sonication are necessary in
the preparation of cell extracts for the assay of
cellular enzymes. In the present study it was found
that both fibroblasts HGPRT and APRT are sensitive to
these treatments and that the resulting inactivation can
be partially prevented by the addition of PRPP.

Cultured human skin fibroblasts, harvested by brief
trypsinization, were suspended to various cell concenent-
ration in10 mM Tris HCl buffer pH 7.4 and extracted by
thrice rapid freezing and thawing in dry ice and aceton
or by sonication. The protecting effect of PRPP and other
compounds was studied by their addition to the cell
suspension prior to the extraction procedures. HGPRT and

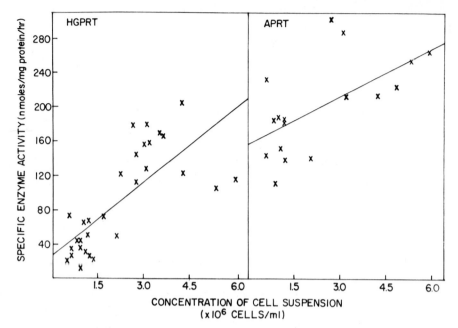

FIG. 1. Sensitivity of fibroblast HGPRT and APRT to inactivation by freezing and thawing. Dependence on cell concentration.

TABLE 1. DEPENDENCE OF PHOSPHORIBOSYLTRANSFERASE INACTIVATION BY FREEZING AND THAWING ON CELL CONCENTRATION AND PRESENCE OF PRPP

Subjects	Specific Activities (nmoles/mg protein/hr)							
	HGPRT				APRT			
	$3x10^{6**}$		$1x10^{6**}$		$3x10^{6**}$		$1x10^{6**}$	
	a	b	a	b	a	b	a	b
NORMAL*	153.5	162	42.5	107	252	260	163	216
HGPRT-DEFICIENT								
LNS	3.6	3.8	1.3	2.8	198	200	152	175
GOUT	14.6	13.8	6.2	6.3	250	216	169	248

a   Without PRPP
b   With 1 mM PRPP
*   Mean of 15 determinations on 5 subjects
** Cells/ml

APRT were assayed in the dialyzed extracts by a radiochemical assay (6).

Freezing and thawing of cell suspensions containing
less than 6.0×10^6 cells/ml caused inactivation of both
HGPRT and APRT, depending on the cell concentration
(Fig. 1), the sensitivity of both cellular phosphoribosyltransferases increasing with cell dilution. The rate of
increase in sensitivity to freezing and thawing by cell
dilution was greater for HGPRT than for APRT; 3-fold
dilution from 3×10^6 to 1×10^6 cells/ml resulted in a
decrease in the specific activities of HGPRT and APRT
by 75% and 34%, respectively, the ratio of the specific
activity of APRT to that of HGPRT increasing from 1.6
to 3.8.

Addition of bovine serum albumin (0.6 mg/ml) or
mercaptoethanol (2.5 mM) to the dilute cell suspensions
prior to freezing and thawing did not protect the
cellular phosphoribosyltransferases from inactivation.
On the other hand, PRPP in concentrations as low as
0.05 mM partially stabilized both enzymes against the
inactivation; in the presence of 1 mM PRPP the decrease
in the specific activities of HGPRT and APRT by the above
3-fold dilution was 30% and 14% respectively, causing
the ratio of the specific activities of APRT and HGPRT
to increase from 1.6 to 2.0 only.

Extracts of cultured fibroblasts obtained from two
HGPRT-deficient subjects, one with LNS, the other with
partial enzyme deficiency associated with gout, were also
studied. The APRT in the fibroblasts of both these
patients and the HGPRT in the LNS cells behaved similarly
to the phosphoribosyltransferases of the normal subjects,
both in regard to inactivation by freezing and thawing
and to protection by PRPP. On the other hand the HGPRT
in the fibroblasts of the gouty subject exhibited normal
sensitivity to inactivation by freezing and thawing but
the protective effect of PRPP was minimal (Table 1).

PRPP has been previously shown to protect HGPRT and
APRT against thermal inactivation in-vitro (7). In addition an increased concentration of PRPP in LNS erythrocytes has been implicated in the stabilization of the
APRT in such cells against degradation in vivo (7).
Mutant HGPRT enzymes, have been found to exhibit altered
sensitivity to thermal inactivation (8) and to protection
against such treatment by PRPP (9). Our finding that a

mutant HGPRT could not be protected normally by PRPP
against inactivation by freezing and thawing is another
reflection of the structural mutation of this enzyme.

The finding that HGPRT and APRT are partially inacti-
vated by freezing and thawing is also of importance
regarding the determination of the specific activities
of these enzymes in extracts of cultured fibroblasts.
Freezing and thawing is a common step in preparation of
cell extracts and the concentration of cell suspensions
exhibiting sensitivity to inactivation is within the
range used in assays. In view of the results reported,
care should be taken to perform the freezing and thawing
step on cell suspensions containing more than $6x10^6$ cells/
ml and to protect the enzymes by the addition of PRPP.
The same considerations hold for assays of cell extracts
prepared by sonication. Similar inactivation of the
phosphoribosyltransferases by sonication could also be
protected against by PRPP.

References

1. Kornberg A., Lieberman I. and Simms E.S., J. Biol.
   Chem. 215:417, 1955.

2. Hori M. and Henderson F.J. J. Biol. Chem. 241:3404,
   1966.

3. Seegmiller J.E., Rosenbloom F.M. and Kelley W.N.
   Science 155:1682, 1967.

4. Lesch M. and Myhan W.L. Amer. J. Med. 36:561, 1964.

5. Kelley W.N., Rosenbloom F.M., Henderson J.F. and
   Seegmiller J.E. Proc. Nat. Acad. Sci. U.S.A.
   57:1735, 1967.

6. Sperling O., Frank M., Ophir R., Liberman U.A.,
   Adam A. and De Vries A. Europ. J. Clin. Biol. Res.
   15:942, 1970.

7. Greene M.L., Bayles J.R. and Seegmiller J.E.
   Science 167:887, 1970.

8. Kelley W.N. and Meade J.C. J. Biol. Chem.
   246:2953, 1971.

9. McDonald J.A. and Kelley W.N. Science 171:689, 1971.

PURIFICATION AND CHARACTERIZATION OF A NEURAL HYPO-
XANTHINE-GUANINE-PHOSPHORIBOSYLTRANSFERASE (HGPRT)

Wolf Gutensohn

Max-Planck-Institut für Biochemie

D 8033 Martinsried, Munich, Germany

Starting from investigations on RNA-metabolism in rat brain and because of the close connection of HGPRT with a neurological disorder in man we became interested in the properties of a neural HGPRT in the laboratory of Gordon Guroff at the National Institute of Health. We adjusted the known HGPRT-assay using DOWEX-1 X 8 for separation of the product to smaller scale and used this assay (W.Gutensohn,G.Guroff, Analytical Biochemistry 47,132(1972)) in all the following experiments.

Starting from a crude supernatant of rat brain a partial purification was achieved by ammonium sulfate fractionation, a heat treatment and DEAE-cellulose chromatography. A pattern of at least 3 isoenzyme-peaks (designated fraction I,II, and III) becomes apparent, but the ratio of activity with the two possible substrates (guanine and hypoxanthine) remains constant over the whole range. On polyacrylamid gel electrophoresis (according to Davis and Ornstein) activity from all three isoenzyme-fractions migrates as one sharp band to the same position. In the most highly purified fraction I enzyme activity was enriched about 650-fold over the starting material and as judged by the pattern on polyacrylamide gel this preparation was about 50% pure. It was used for some kinetic and inhibition studies.

Kinetics for guanine are of a complicated type and appear to be sigmoid. Kinetics for phosphoribosyl-

pyrophosphate (PRPP) are linear and give a K_M of
o.2 mM. The product GMP and a number of other nucleotides inhibit the enzyme. An interesting observation
is, that with the nucleoside phosphates of the substrate guanine inhibition decreases with increasing
number of phosphates, whereas with all the other bases
(adenin and the pyrimidines) inhibition increases with
increasing number of phosphates. Especially with the
triphosphates inhibition might be partly due to competition for the magnesium ion, but it cannot be completely abolished by an excess of Mg^{++}. All the nucleotides we have looked at – with the possible exception
of IMP – are competitive inhibitors with respect to
PRPP. Approximate K_i-values are: GMP 0.02 mM, GDP
0.09 mM, GTP 0.15 mM, AMP 3.4 mM, ATP 1.6 mM, CMP
1.3 mM, and UMP 6 mM.

A number of polyacrylamide gel electrophoreses
from fraction I were run and not fixed and stained. At
the position of the enzyme peak in parallel gels appropriate slices were cut out, macerated, mixed with
Freund's adjuvant and used to immunize a rabbit. The
antiserum obtained had a low titer but inhibited the
activity of all 3 isoenzymes. It did not crossreact
with HGPRT from rabbit or human erythrocytes.

A more highly purified enzyme was obtained by
adding Sephadex gel filtration and preparative isoelectric focusing to the above mentioned purification
steps. Thus in one case a pure enzyme was obtained, as
judged by polyacrylamide gel electrophoresis, although
in extremely low yield. The molecular weight as determined by gel filtration on a Sephadex G 100 column
calibrated by standard proteins was 63 000 and 64 000
in two separate runs. SDS-gel electrophoresis gave a
subunit molecular weight of 26 000 which appears to be
rather low, if one assumes that the enzyme contains
two identical subunits. Preparative isoelectric focusing on a pH-gradient of pH 3 – 6 did not give a good
resolution of isoenzyme peaks. Using a gradient of
pH 4 – 6 lead to two major peaks with a pI around 5.6
and 6.3 respectively.

As already mentioned for the purification procedure the enzyme is heat stable. An ammonium sulfate
fraction when heated to 65°C retains its activity
during 7 min without any significant decrease. On the
other hand storage of the purified enzyme is always a
problem. As yet no satisfactory way of storage has
been found. Rapid loss of activity occurs at +4°C as

well as at -18°C. High concentrations of mercaptoethanol (0.1M) are rather detrimental than helpful. The
best way to protect the enzyme is addition of bovine
serum albumin at 1 mg/ml. But even under these conditions the enzyme loses over 80% of its activity within 3 months. Serum albumin is also used to stabilize
HGPRT in very dilute solutions during assays.

In summary it can be said that except for minor
differences the neural enzyme described here is very
similar in its physical and kinetic properties to the
human erythrocyte HGPRT.

After cell fractionation of rat brain the highest
specific as well as the major part of the total HGPRT
activity is found in the soluble supernatant fraction,
as expected. However, when a crude mitochondrial fraction is washed to constant activity and then subfrac-
tionated on a discontinuous Ficoll-gradient, activity
can be enriched in synaptosomes. Whether the enzyme
is only entrapped there as a normal constituent of the
cytoplasm during the disruption of the nerve endings
or whether it serves a more specific function there,
remains to be determined.

Earlier studies by Rosenbloom and coworkers on
the HGPRT levels in different areas of the normal human brain have demonstrated very low levels in spinal
cord, intermediate levels in cortex and cerebellum and
high levels in the basal ganglia, which corresponds
well with the neurological symptoms observed in the
Lesch-Nyhan condition. In rat brain the distribution
of the enzyme is more homogeneous. Levels in spinal
cord and medulla are slightly lower, but activities in
cerebellum, diencephalon, basal ganglia and cortex lie
within a rather narrow range.

The development of the enzyme activity in whole
brains of the young rat shows a sharp rise from one
day before birth to about the 15th day followed by a
plateau up to at least 6 months. This increase goes
parallel to the most extensive general development and
myelination of the rat brain.

In a first attempt to determine the turnover of
HGPRT in vivo rather high doses of cycloheximide
(10 mg/kg) were given to young rats up to 7 times in
1 hour intervals. Control animals received saline in-
jections. At intervals animals were sacrificed and
HGPRT activity was assayed in brain, liver and kid-

neys. Although the drug doses were high enough to kill
the remaining animals overnight there was no significant decrease of enzyme activity in any of the three
organs within this short period. This experiment will
have to be repeated with lower, less toxic doses of
protein biosynthesis inhibitors to allow for a longer
period to observe enzyme activities.

HUMAN ADENINE PHOSPHORIBOSYLTRANSFERASE: PURIFICATION, SUBUNIT

STRUCTURE AND SUBSTRATE SPECIFICITY

C. B. Thomas, W. J. Arnold and W. N. Kelley

Department of Medicine, Duke University Medical Center

Durham, North Carolina 27710

Adenine phosphoribosyltransferase (APRT) catalyzes the magnesium dependent conversion of adenine to adenosine 5'-monophosphate (AMP) utilizing the high energy compound, 5-phosphoribosyl-1-pyrophosphate (PP-ribose-P) as a cosubstrate. In humans this enzyme provides the only apparent pathway for conversion of dietary adenine into utilizable nucleotides. The finding of elevated levels of APRT activity in patients with the Lesch-Nyhan syndrome (Seegmiller, Rosenbloom and Kelley, 1967; Kelley, 1968) and the discovery of at least three families with a genetically determined partial deficiency of APRT activity in circulating erythrocytes (Kelley, et al., 1968; Kelley, Fox and Wyngaarden, 1970; Emmerson, et al., present symposium) stimulated our study of a highly purified preparation of human adenine phosphoribosyltransferase.

Erythrocytes from four units of outdated whole blood were washed with saline, lysed by freeze-thawing, and applied to a DEAE-cellulose column at pH 7.0. A single peak of APRT activity was eluted with a 0 to 0.3 M potassium chloride gradient. This was subjected to a 35% to 60% ammonium sulfate fractionation followed by a 30% ethanol precipitation at pH 5.0. The APRT in the resuspended precipitate was chromatographed on CM-cellulose at pH 5.2, concentrated by hyperosmolar dialysis and applied to a Sephadex G-75 column. The activity peak was pooled and used for further studies.

A major problem in the APRT purification was stabilization of the enzyme. While APRT activity in crude hemolysate is relatively stable, enzyme activity rapidly deteriorates at 4^0 in more purified preparations, a finding noted by previous investigators (Hori and Henderson, 1966; Srivastava and Beutler, 1971). Stability studies on

TABLE 1

PURIFICATION OF ADENINE PHOSPHORIBOSYLTRANSFERASE FROM HUMAN ERYTHROCYTES

Step	Volume	Specific Activity	Total protein	Recovery	Purification
	ml	umoles/mg/min	mg	% initial activity	-fold
Hemolysate	562	0.000287	360,000	100	
DEAE-eluate	397	0.0380	1,500	55	130
Ammonium sulfate precipitation	9.3	0.0375	460	17	130
Ethanol precipitation	8.5	0.0790	180	14	280
CM-cellulose column	7.3	9.58	1.0	9.5	33,000
Osmotic concentration	1.0			7.8	
Sephadex G-75	7.1	5.08	0.54	2.7	18,000

(From Thomas, Arnold and Kelley, 1973)

partially purified APRT revealed that dithiothreitol, dimethylsulfoxide, and ammonium sulfate each provide partial stabilization of the
enzyme, but only PP-ribose-P gives complete protection. For this
reason 5 mM Mg^{++} and 0.1 mM PP-ribose-P were added to all buffers
used in the latter stages of the purification.

    Table I summarizes the purification. After the CM-cellulose
chromatography, the enzyme had a specific activity of 9.58 umoles
of AMP formed per mg protein per minute, which represents a 33,000
fold purification from hemolysate. The final gel chromatography
produced a two-fold further protein purification, although specific
activity decreased in this highly purified preparation due to enzyme
instability even in the presence of Mg^{++} and PP-ribose-P.

    When the purity of the final preparation was examined by disc
gel electrophoresis, there was one major band which was shown
by gel scanning to comprise 80% of the protein present. APRT was
assayable from this band (Fig. 1).

    To define the subunit structure of human adenine phosphori-

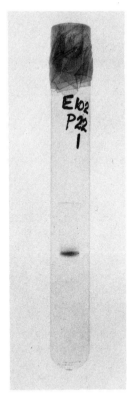

Fig. 1. Disc gel electrophoresis of highly purified human adenine
phosphoribosyltransferase (From Thomas, Arnold and Kelley, 1973).

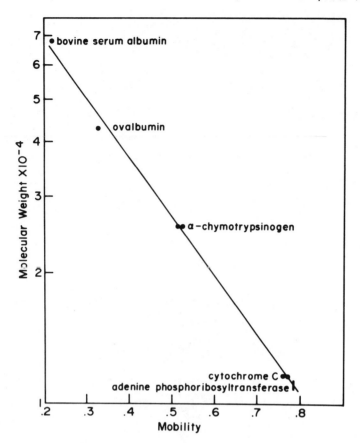

Fig. 2. Subunit molecular weight of human adenine phosphoribosyltransferase (From Thomas, Arnold and Kelley, 1973).

bosyltransferase a sample of the most highly purified enzyme
preparation was used for sodium dodecyl sulfate electrophoresis.
This system demonstrated a single major protein band with a
mobility which, when compared to that for known protein standards,
indicated a subunit molecular weight of 11,100 (Fig. 2). Omission
of β-mercaptoethanol from the samples and gels gave similar results
which suggest that the APRT subunits are not bound by disulfide
bridges.

    A 280-fold purified enzyme preparation was mixed with α-chymo-
trypsinogen and bovine serum albumin and subjected to sucrose
density gradient ultracentrifugation (Fig. 3). Comparison of the
sedimentation distance for APRT with that for the protein standards
assigned the enzyme an $S_{20,w}$ of 3.35.

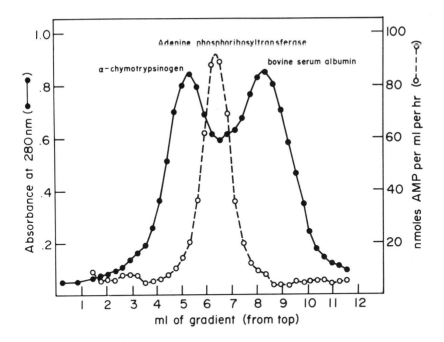

Fig. 3. Sucrose density gradient ultracentrifugation of human adenine phosphoribosyltransferase (From Thomas, Arnold and Kelley, 1973).

APRT and protein standards were chromatographed on Sephadex G-100 and G-75 columns. The elution volumes were converted to K_{av}'s and from these the inverse error function complements of the column partition coeffients were derived and plotted against Stokes radius (Fig. 4). Two such determinations assigned APRT a mean Stokes radius of 24.9 Å. Applying the method of Siegel and Monty (1966) one calculates a molecular weight of 34,400 and frictional coefficient of 1.16 from the Stokes radius and sedimentation coefficient. Alternatively the molecular weight of APRT as determined from the mean of the values from four gel chromatography experiments was 34,000.

While APRT activity is usually eluted from gel filtration columns in a single peak of activity, on two occasions this was not the case. Figure 5 shows the results from one such occasion, with enzyme activity plotted as a function of volume relative to the elution volume of bovine serum albumin. The APRT activity maxima correspond to molecular weights of 69,000 and 31,000. Another experiment revealed activity peaks corresponding to molecular

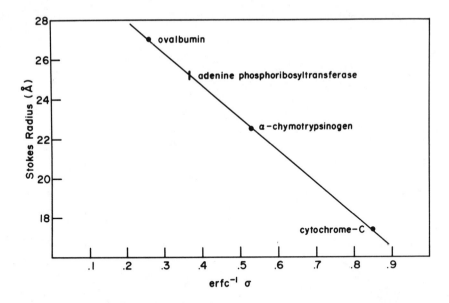

Fig. 4. Stokes radius of human adenine phosphoribosyltransferase
(From Thomas, Arnold and Kelley, 1973).

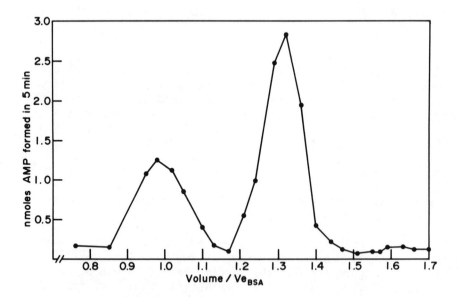

Fig. 5. Gel filtration of human adenine phosphoribosyltransferase
on Sephadex G-100. (From Thomas, Arnold and Kelley, 1973).

weights of 34,500 and 21,000. We have interpreted these molecular
weight forms as dimer, trimer and hexamer forms of the basic 11,000
molecular weight subunit.   No characteristic conditions were noted
which would reproducibly generate these alternate aggregation states,
and under most conditions the enzyme appears to exist in the trimer
form. When enzyme activity was studied as a function of pH, APRT
was found to be maximally active over a wide pH range from 7.4 to
9.5. Figure 6 illustrates the behavior of APRT upon isoelectric
focusing of a partially purified preparation. Activity, pH, and
absorbance at 280 nm are plotted in relation to the elution fraction
from the focused column. APRT was eluted as a single peak of
activity with an isoelectric point of 4.85 and 4.70 on two separate
determinations.

     The most highly purified enzyme preparation was studied for
substrate specificity (Table 2). The results are expressed as
per cent of nucleotide formed from each substrate as compared to
adenine. Adenine phosphoribosyltransferase shows substantial
activity toward 4-amino-5-imidazolecarboxamide and 2,6-diaminopurine
but not hypoxanthine, guanine, 6-mercaptopurine, or adenosine.

     Studying APRT from rat liver, Groth and Young (1971) have

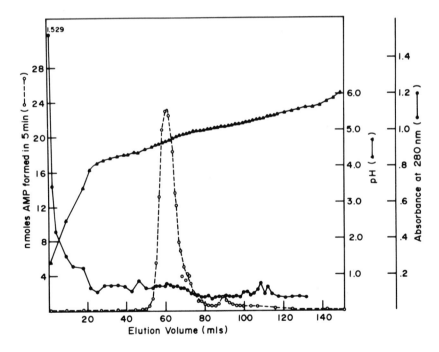

Fig. 6.   Isoelectric focusing of human adenine phosphoribosyltransferase.

C. B. THOMAS, W. J. ARNOLD, AND W. N. KELLEY

TABLE 2

SPECIFICITY OF HUMAN ADENINE PHOSPHORIBOSYLTRANSFERASE FOR PURINE
BASES AND NUCLEOSIDES

Substrate	Concentration	Nucleotide formed/ ug protein
	mM	%
Adenine	0.30	100
Hypoxanthine	0.50	<0.1
Guanine	0.25	<0.1
4-Amino-5-imidazolecarboxamide	2.0	11.8
2,6-Diaminopurine	1.0	10.8
6-Mercaptopurine	0.40	0.13
Adenosine	0.77	<0.1

(From Thomas, Arnold and Kelley, 1973)

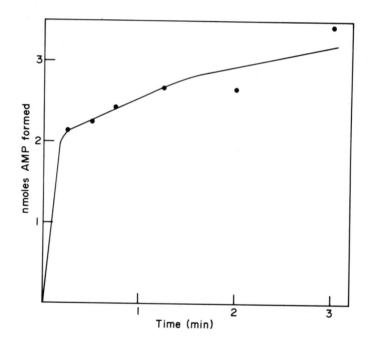

Fig. 7. Burst of AMP synthesis by partially purified human adenine
phosphoribosyltransferase (From Thomas, Arnold and Kelley, 1973).

previously described a rapid and limited synthesis of AMP at 0^0 in the presence of substrates but in the absence of divalent cation. They hypothesized that PP-ribose-P binds to the enzyme causing a conformational change which exposes a bound molecule of ribose-5-phosphate for rapid reaction with adenine. Our investigation of the highly purified APRT revealed a similar phenomenon (Fig. 7). There was a measurable amount of AMP formation at the earliest time point after substrates were added as compared to the control reaction which yielded no AMP when adenine but not PP-ribose-P was added. Moreover we found that an enzyme preparation subjected to these conditions could again produce a burst of AMP synthesis if it were first incubated at 37^0 with Mg^{++}, PP-ribose-P, and adenine to permit the presumed replacement of a bound ribose-5-phosphate group.

REFERENCES

Emmerson, B.J., Gordon, R.B. and Thompson, L. 1973. Adenine phosphoribosyltransferase deficiency in a female with gout. Present symposium.

Groth, D.P. and Young, L.G. 1971. On the formation of an intermediate in the adenine phosphoribosyltransferase reaction. Biochem. Biophys. Res. Commun. 43: 82-87.

Hori, M. and Henderson, J.F. 1966. Purification and properties of adenylate pyrophosphorylase from Ehrlich Ascites Tumor cells. J. Biol. Chem. 241: 1406-1411.

Kelley, W.N. 1968. Hypoxanthine-guanine phosphoribosyltransferase deficiency in the Lesch-Nyhan syndrome and gout. Fed. Proc. 27: 1047-1052.

Kelley, W.N., Levy, R.I., Rosenbloom, F.M., Henderson, J.F. and Seegmiller, J.E. 1968. Adenine phosphoribosyltransferase deficiency: A previously undescribed genetic defect in man. J. Clin. Invest. 47: 2281-2289.

Kelley, W.N., Fox, I.H. and Wyngaarden, J.B. 1970. Further evaluation of adenine phosphoribosyltransferase deficiency in man: Occurrence in a patient with gout. Clin. Res. 18: 53.

Seegmiller, J.E., Rosenbloom, F.M. and Kelley, W.N. 1967. Enzyme defect associated with a sex-linked human neurological disorder and excessive purine synthesis. Science 155: 1682-1684.

Siegel, L.M. and Monty, K.J.   1966.   Determination of molecular
    weights and frictional ratios of proteins in impure systems by
    use of gel filtration and density gradient ultracentrifugation.
    Application to crude preparations of sulfite and hydroxylamine
    reductase.   Biochem. Biophys. Acta 112: 346-362.

Srivastava, S.K. and Beutler, E.   1971.   Purification and kinetic
    studies of adenine phosphoribosyltransferase from human
    erythrocytes.   Arch. Biochem. Biophys. 142: 426-434.

Thomas, C.B., Arnold, W.J. and Kelley, W.N.   1973.   Human adenine
    phosphoribosyltransferase: Purification, subunit structure,
    and substrate specificity.   J. Biol. Chem. 248: 2529-2535.

GENETIC CONTROL OF BACTERIAL PURINE PHOSPHORIBOSYL-
TRANSFERASES AND AN APPROACH TO GENE ENRICHMENT

Joseph S. Gots and Charles E. Benson

University of Pennsylvania

Philadelphia, Pennsylvania 19174, USA

Enteric bacteria such as Escherichia coli and Salmonella typhimurium possess three separate genes that control three different phosphoribosyl-transferases (PRT). A-PRT (apt gene) is active only for adenine; G-PRT (gpt gene) converts guanine, hypoxanthine and xanthine; and H-PRT (hpt gene) is primarily active for hypoxanthine, poorly for guanine and not at all for xanthine.

Biochemical and genetic analyses of mutants that are deficient in these enzymes allows for the assignment of function in terms of percent of total conversion of the individual bases to their respective ribonucleotides. Table I summarizes the responsibilities of each of the enzymes in the total con-versions. A-PRT is known from our previous studies to be exclusively in-volved in the total conversion of adenine (1). This paper will deal with an analysis of the gpt and hpt genes and their products, G-PRT and H-PRT, respectively.

Table I: Summary of activities of purine phosphoribosyltransferases in Salmonella typhimurium.

Enzyme	Gene	Percent total conversion of			
		adenine	hypoxanthine	guanine	xanthine
A-PRT	apt	100	0	0	0
G-PRT	gpt	0	25-35	90	100
H-PRT	hpt	0	65-75	10	0

The gpt gene was first detected in Salmonella among deletion mutations involving the proAB region (2). Some of the mutants with chromosomal deletions of this region were simultaneously deleted for gpt activity. The phenotype was particularly expressed in purine requiring auxotrophs as an inability to use xanthine or guanine as a purine source for growth. Growth on hypoxanthine and adenine was normal. In the absence of a purine requirement the gpt mutation is expressed as resistance to the inhibitory action of 8-azaguanine.

The hpt gene was discovered by Chou and Martin (3) in Salmonella. The phenotype can be detected in a pur⁻ mutant which in addition is unable to convert IMP to GMP (e.g. guaB lacking IMP dehydrogenase) and also unable to convert GMP to IMP (guaC⁻ GMP reductase). Such a mutant, pur⁻hpt⁻guaB⁻guaC⁻, will grow only when given adenine plus guanine. Strain TM445 shown in the Tables is a guaB⁺ derivative of this multiply marked mutant. It grows on hypoxanthine, presumably by using the G-PRT, but this growth is drastically inhibited by guanine which inhibits the conversion of hypoxanthine by G-PRT. An hpt⁻gpt⁻ double mutant was obtained from strain TM445 by transduction techniques. This double mutant grows only on adenine, but poorly. Growth on adenine can be stimulated by GMP or 2,6-diamino-purine which is converted by A-PRT and then deaminated to GMP.

Table II shows the PRT-ase activities of the various mutants expressed as percent of wild-type activity. The hpt mutant is primarily affected in its IMP activity. The remaining activity is due to G-PRT and can be completely inhibited by the addition of guanine to the reaction mixture (data not shown). The gpt mutant is deficient in GMP and XMP activity with only minimal decrease in IMP activity. The residual activity for XMP in all gpt strains is an artifact imposed by the rapid assay using DE-81 Whatman discs to capture the nucleotide. The bases are usually washed off by NH_4HCO_3 and water, but this washing does not allow complete removal of the C^{14} xanthine substrate. In a more tedious method involving separation of base and nucleotide by thin layer chromatography, XMP activity in the gpt strains is virtually undetectable.

Table II also shows the complete recovery of GMP and XMP in the merodiploid strain PRO47/F'. This strain carries an episome which contains the gpt⁺ allele, and is genotypically gpt⁻/gpt⁺. As would be expected, the double mutant (hpt⁻gpt⁻) is deficient in all activities. The strains with the "sug" designation (GP-36 and GP-55) were isolated from gpt⁻ strains as a secondary (suppressor) mutation which restored the ability to grow in guanine without allowing growth in xanthine. Restoration of GMP activity in such a mutant is shown in the Table, but note also that this occurs at the

Table II: PRTase activities of representative mutants

Strain	Genotype	Phosphoribosyltransferase Activity (% of wildtype)		
		IMP	GMP	XMP
LT2	$hpt^+ gpt^+$	100	100	100
TM445	$hpt^- gpt^+$	23.6	98.5	92.9
PRO47	$hpt^+ gpt^-$	85.3	8.4	7.3
PRO47/F[1]	$hpt^+ gpt^-/gpt^+$	82.1	102.2	98.3
GP-36	$hpt^+ (sug) gpt^-$	47.0	50.3	8.9
GP-54	$hpt^- gpt^-$	1.2	7.6	6.8
GP-55	$hpt^- (sug) gpt^-$	21.1	33.1	6.7

Specific activities of wildtype (100%) = IMP (33.9), GMP (20.2), XMP (28.0)

expense of IMP activity (compare GP-36 with PRO47). Evidence indicates that the sug mutation is a mutation in the hpt gene that creates a modified H-PRT enzyme with increased efficiency for the utilization of guanine. Further details supporting this will be given below.

The existence of two separate enzymes for the conversions of hypoxanthine, guanine and xanthine was revealed primarily by the genetic studies described above. Additional biochemical evidence was obtained by separation of activities in cell-free extracts prepared from the wild type parental strain. This was achieved by column chromatography using ECTEOLA cellulose. Figure 1 shows the elution pattern from such an experiment. Results similar to these have been reported for E. coli strains (4). Distribution of IMP activity is compatible with the distribution suggested by the mutant strains. Most of this activity is in the first peak and a smaller amount is in the second. The GMP activity is considerably less than the IMP activity in the first peak, but approximately equal to it in the second. Note however, that the total GMP activity is approximately evenly distributed between the two peaks. This is not what would have been expected from the mutant activities which suggested a 90-10 distribution of guanine activity between G-PRT and H-PRT, respectively. From this, it appears

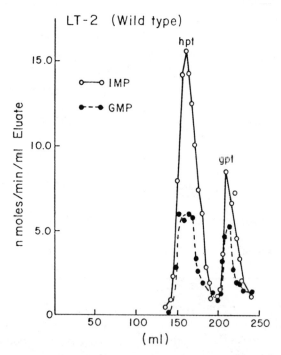

Figure 1: ECTEOLA cellulose column chromatography of PRTase activities of wild type Salmonella strain.

100 mg of extract protein was applied to column (1 x 22 cm) and eluted with 9 mM Tris–Cl, 0.1 mM $MgSO_4$ buffer at pH 7.4 using linear gradient between 0 and 0.125 M KCl. Flow rate was 15 ml/hr; 3–4 ml fractions collected. Assayed with ^{14}C substrate and nucleotide formed was measured in 0.025 ml of reaction mixture applied to DE–81 Whatman discs. Discs were washed (30 min) in 4 mM NH_4HCO_3 and two times (60 min each) in distilled water; dried and counted in scintillation counter.

possible that the gpt product is required for optimal guanine activity of the hpt product.

The identification of the two peaks in terms of gene products was made possible by similar experiments using extracts from the various mutant strains. Table III summarizes the results which show that peak I is the product of the hpt gene and peak II is the product of the gpt gene. Note that in the gpt⁻ mutant the hpt product (peak I) shows no significant GMP activity. This further indicates that the gpt product is required for the demonstrable GMP activity found in peak I in the parental strain. Note also that GMP activity is restored in the gpt mutant by the secondary sug mutation and that this activity is coincidental with IMP activity. The details are presented in Figure 2.

This Figure compares the elution pattern of the original gpt mutant with that of its sug derivative. In both cases, only one peak was found and this corresponded in location with the hpt product (peak I). The sug mutation led to a reduction of specific activity for hypoxanthine. Guanine activity was undectable in the gpt mutant but was restored to a level equal to that of hypoxanthine activity in the sug derivative.

We postulate that the sug mutation leads to a genetically modified H–PRT due to a mutation in the hpt gene and that this modification creates a change in substrate specificity by allowing for a more efficient utilization of guanine. This idea would predict that 1) guanine and hypoxanthine

---

Table III: Chromatographic elution patterns of various mutant strains.

Strain	Genotype	Activity Elution Patterns	
		Peak I (hpt) (130–170 ml)	Peak II (gpt) (190–220 ml)
LT-2	hpt⁺ gpt⁺	IMP-GMP	IMP-GMP-XMP
TM445	hpt⁻ gpt⁺	IMP	------
PRO47	hpt⁺ gpt⁻	-----	IMP-GMP-XMP
PRO47/F¹	hpt⁺ gpt⁻/gpt⁺	IMP-GMP	IMP-GMP-XMP
GP-36	hpt (sug) gpt⁻	IMP-GMP	------

activities should be inseparable on chromatographic elution, 2) both
activities should be found in the peak corresponding to the hpt product
(H–PRT), and 3) sug should be an allelic form of hpt. These first two predictions were realized in the data presented in Table III and Figure 2. The
third prediction was verified in a genetic cross between a gpt⁻sug donor and
a hpt⁻gpt⁻ recipient. Selection was made for the ability to grow on hypoxanthine (hpt⁺ phenotype) and all recombinants obtained were found to have
the donor phenotype (gpt⁻sug) in that they could also grow on guanine, but
not xanthine. No recombinants with the hpt⁺gpt⁻ phenotype (growth on
hypoxanthine but not guanine) were obtained. This indicates a 100%
linkage between hpt and sug and hence allelic identity.

The close proximity of the gpt gene to the pro region places it close to
the attachment site of the P22 lysogenic phage. This presents an unusual

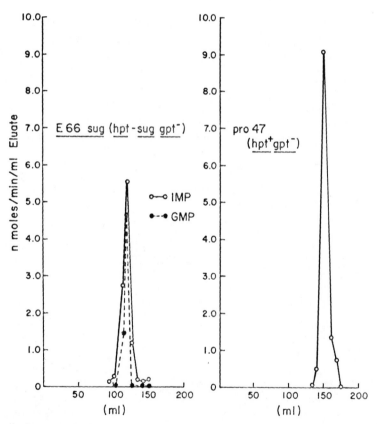

Figure 2: Comparison of elution patterns of a gpt mutant (PRO47) and
its sug derivative (E66sug).

opportunity to obtain DNA preparations that are especially enriched for the gpt gene. It should be possible to isolate specialized transducing particles following induction of a P22 prophage from a lysogenic strain. Such particles carrying the pro genes have been obtained by Jessop (5). We have examined some of Jessop's lysogens, have confirmed that the phages have specialized transducing properties for the pro genes, but in no case have we found that they also carry gpt. Another approach was based on the observation by Roth (6) that the E. coli episome, F'prolac, also has a P22 attachment site, and when present in Salmonella, the episome can be lysogenized with P22 phage. We have created such a lysogen and upon induction to yield active phage, the lysate was found to contain rare specialized co-transducing particles for both gpt and pro. Relatively high frequency transducing lysates have been obtained from gpt⁺ transductants and the properties of these lysates may be summarized as follows:

    1. The transducing efficiency for gpt and pro is 10^{-2} (transductants per phage) which represents a 10^4 fold increase over that of a generalized transducing lysate (10^{-6}).

    2. Transducing particles can be propagated through the lytic cycle in a gpt⁻ host.

    3. The transductants obtained are relatively unstable and segregate to gpt⁻.

    4. Cesium chloride density gradients show identity between transducing particles and phage at a density of 1.528.

    5. Both phage and transducing activities are resistant to DNase.

    6. P22 phage antibody neutralizes phage activity more efficiently than transducing activity. This is also true for inactivation by ultraviolet irradiation.

References

1. Kalle, G. P. and Gots, J. S. Science, 142: 680 (1963).

2. Gots, J. S., Benson, C. E., and Shumas, S. R. J. Bacteriol., 112: 910 (1972).

3. Chou, J. Y. and Martin, R. G. J. Bacteriol., 112: 1010 (1972).

4. Krenitsky, T. A., Neil, S. M., and Miller, R. L. J. Biol. Chem., 245: 2605 (1970).

5. Jessop, A. P. Mol. Genl. Genet., 114: 214 (1972).

6. Roth, J. R. and Hoppe, I. Genetics, 71: 53 (1972).

Glutamine-PRPP Amidotransferase

HUMAN GLUTAMINE PHOSPHORIBOSYLPYROPHOSPHATE (PP-RIBOSE-P) AMIDOTRANS-

FERASE: KINETIC, REGULATION AND CONFIGURATIONAL CHANGES

E. W. Holmes, Jr., J. B. Wyngaarden and W. N. Kelley

Duke University Medical Center, Durham, North Carolina

27710

In a large percentage of patients with gout, hyperuricemia is the result of an increase in the rate of purine biosynthesis de novo (Wyngaarden and Kelley, 1972). Consequently, it is important to understand the molecular basis for the regulation of purine biosynthesis in man.

Figure 1 depicts the important steps in purine biosynthesis de novo. The first reaction which is unique to this pathway is catalyzed by the enzyme glutamine phosphoribosylpyrophosphate amidotransferase. The substrates for this reaction are phosphoribosylpyrophosphate, or PP-ribose-P, glutamine and water and the products are phosphoribosylamine, glutamate and inorganic pyrophosphate. Data obtained from the study of many species indicate that this reaction catalyzed by amidotransferase is the rate limiting step in purine biosynthesis. Phosphoribosylamine is then converted by a series of enzymatic reactions to the parent purine ribonucleotide IMP, and IMP is subsequently converted to GMP and AMP.

In man there is convincing evidence that purine biosynthesis de novo is controlled by the intracellular concentrations of PP-ribose-P and purine nucleotides. When PP-ribose-P levels are increased or decreased there is a corresponding change in the rate of purine biosynthesis (Kelley, Fox and Wyngaarden, 1970; Kelley, et al., 1970; Fox and Kelley, 1971) suggesting that amidotransferase is not saturated with this substrate in vivo. We have determined the apparent Km for PP-ribose-P of the normal human enzyme to be 5×10^{-4} Molar, a value which is 10 to 100 times greater than the intracellular concentration of PP-ribose-P (Fox and Kelley, 1971). In addition to this regulatory process, the administration of purine bases in vivo or in tissue culture leads to a decrease in

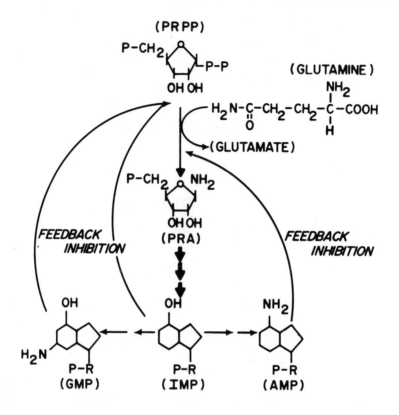

Fig. 1. Important steps in purine biosynthesis de novo.

purine biosynthesis (Fox and Kelley, 1971; Schulman, et al., 1971).
This is due in part to feedback inhibition of amidotransferase by
purine nucleotides (Nierlich and Magasanik,1965; Momose, Nishikawa
and Katsuya, 1965; Nagy, 1970; Wyngaarden and Ashton, 1959; Hartman,
1963; Caskey, Ashton and Wyngaarden, 1964; Rowe and Wyngaarden,
1968; Rowe, Coleman and Wyngaarden, 1970; Hill and Bennett, 1969;
Reem, 1972; Wood and Seegmiller, 1973; Holmes, et al., 1973). Both
of these observations point to the critical role of PP-ribose-P
and purine nucleotides in the regulation of amidotransferase as
well as purine biosynthesis.

The importance of the relative intracellular concentrations
of PP-ribose-P and purine ribonucleotides is illustrated in Fig. 2.

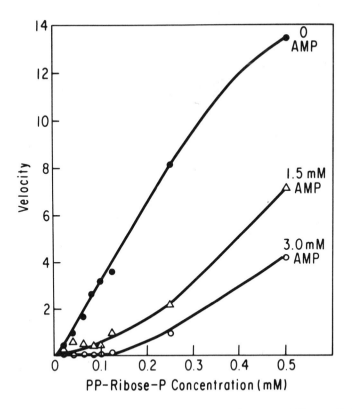

Fig. 2. Michaelis-Menten plot with variable PP-ribose-P in the absence and presence of AMP. Assay performed in 50 mM potassium phosphate buffer, pH 7.4, containing 4 mM glutamine, 5 mM MgCl$_2$, and 18 mM β-mercaptoethanol. Equimolar magnesium was added for each concentration of AMP used. No AMP (●–●), 1.5 mM AMP (△–△), and 3.0 mM AMP (O–O). (From Holmes, et al., 1973).

The enzyme preparation used for these studies was purified approximately 40 fold from human placenta and assayed as previously described (Holmes, et al., 1973). In the absence of purine ribonucleotides human amidotransferase exhibits Michaelis-Menten kinetics for the substrate PP-ribose-P. However, the inclusion of purine ribonucleotides in the assay system results in a qualitative change in the kinetics from a hyperbolic to a sigmoidal function. The result is a marked inhibition of amidotransferase by purine ribonucleotides at physiological concentrations of PP-ribose-P. On the other hand the inhibition produced by purine ribonucleotides

can be reversed by increasing the relative intracellular concentration of PP-ribose-P. The molecular basis for this regulation of
amidotransferase was not explained by these studies, however.

The data presented in Figure 3 suggested a potential mechanism
for the regulation of human amidotransferase. When a partially
purified preparation of the enzyme from human placenta was applied
to an 8% agarose gel filtration column the enzyme activity eluted
in two distinct peaks. The molecular weight of the small form of
the enzyme was calculated to be 133,000 daltons from the Stokes
radius and sedimentation coefficient and the molecular weight of the
large form was 270,000 daltons using similar calculations. Since all
of our data obtained on the human enzyme suggested that it was an
allosteric protein (Holmes, et al., 1973), one explanation for the
molecular heterogeneity of amidotransferase was an association-

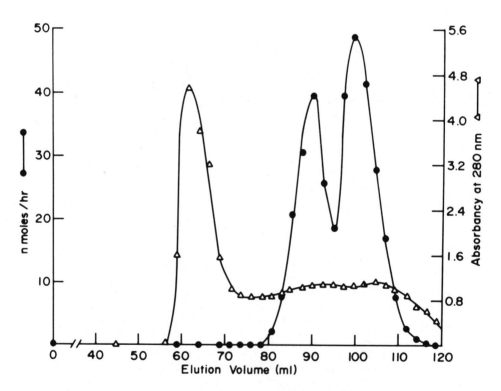

Fig. 3. Molecular heterogeneity of human PP-ribose-P amidotransferase demonstrated by gel filtration. Two milliliters of the enzyme
sample were applied to an 8% agarose column equilibrated with 50
mM potassium phosphate buffer, pH 7.4 containing 5 mM $MgCl_2$, 60 mM
β-mercaptoethanol, and 0.25 M sucrose. (From Holmes, et al., in
press).

dissociation reaction induced by effector molecules. This seemed plausible since the large form of the enzyme was twice the molecular weight of the small form. To explore this in more detail the technique of sucrose gradient ultracentrifugation was employed.

The studies shown in Figure 4 were performed with the small form of the enzyme that had been separated from the large form by gel filtration on Sephadex G-100. When this enzyme preparation was applied to a sucrose gradient that contained no purine ribonucleotide a single peak of amidotransferase activity was observed. The sedimentation coefficient of this peak was 5.9, a value which corresponds to a molecular weight of 133,000. When this same

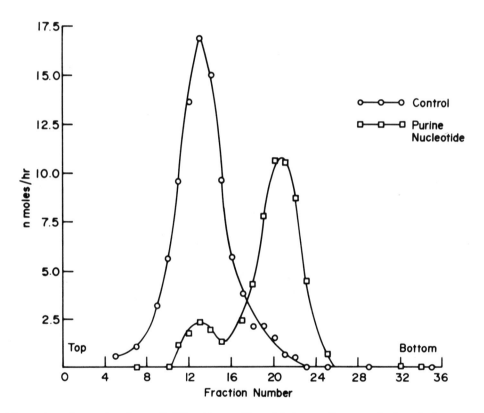

Fig. 4. Association of PP-ribose-P amidotransferase induced by purine ribonucleotides. The small form of the enzyme obtained from Sephadex G-100 column chromatography was incubated at 37° for 15 min. with no purine ribonucleotide (O—O) or 5 mM AMP (□—□) and centrifuged into gradients that contained no purine ribonucleotide or 5 mM AMP, respectively.

enzyme preparation was incubated at 37° with purine ribonucleotide and applied to a gradient that also contained purine ribonucleotide, the major peak of enzyme activity had a sedimentation coefficient of 10.0, a value which corresponds to a molecular weight of 270,000. Thus, the small form of amidotransferase was converted to the large form by purine ribonucleotides.

The large form of the enzyme that had been separated from the small form by gel filtration on 8% agarose was used for the studies shown in Figure 5. When this preparation was applied to a sucrose gradient that contained purine ribonucleo-

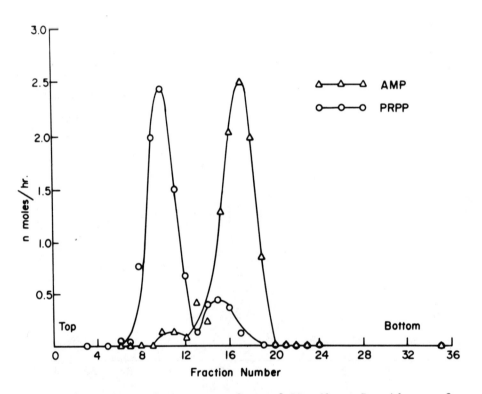

Fig. 5. Conversion of the large form of PP-ribose-P amidotransferase to the small form by PP-ribose-P. The large form of the enzyme was isolated by gel filtration on 8% agarose (Figure 3). This enzyme preparation, containing 5 mM AMP, was incubated at 37° for 15 min. in the absence of PP-ribose-P (△-△) and with 10 mM PP-ribose-P (O—O) and centrifuged into gradients that contained 5 mM AMP (△-△) or 1 mM PP-ribose-P (O—O).

tide, the sedimentation coefficient of amidotransferase was 10.0. If this same preparation was incubated with PP-ribose-P and applied to a gradient that contained PP-ribose-P, the sedimentation coefficient of amidotransferase was 5.9. Therefore, the large form of the enzyme was converted to the small form by PP-ribose-P.

In other experiments L-glutamine alone, with PP-ribose-P or with AMP, was found to have no effect on the molecular size of either the small or large form of human amidotransferase induced by PP-ribose-P or AMP, respectively.

The regulatory significance of the interconversion between these two forms of human amidotransferase was suggested by the following correlation. Purine nucleotides convert the small to the large form of the enzyme and produce complete inhibition of catalytic activity, while PP-ribose-P converts the large to the small form of the enzyme and reverses the inhibition produced by purine ribonucleotides (Holmes, et al., 1973). These observations suggested that the small form of human amidotransferase was the catalytically active species and the large form of the enzyme was inactive. If this hypothesis were correct, it should be possible to correlate enzyme activity with the amount of amidotransferase present as the small form of the enzyme.

For the experiments shown in Figure 6 a partially purified preparation of human amidotransferase was used. In one set of experiments enzyme activity was determined at a non-saturating concentration of PP-ribose-P in the absence of AMP and at several different concentrations of AMP. Enzyme activity is graphed as a percent of maximal activity observed with a saturating concentration of PP-ribose-P in the absence of any purine nucleotide. In the other set of experiments these same assay conditions were reproduced as closely as possible in enzyme samples applied to sucrose density gradients. Using this technique the percentage of amidotransferase present in the small and large form could be approximated under each of the conditions used in the enzyme assays. As shown here enzyme activity decreased as the amount of amidotransferase present in the small form of the enzyme decreased, while the amount of amidotransferase present in the large form increased. These observations provide strong support for an enzyme model in which the small form of human amidotransferase is the active species and the large form of the enzyme is catalytically inactive.

Figure 7 summarizes in schematic fashion the mechanism by which human amidotransferase activity is controlled. PP-ribose-P causes the large inactive form of human amidotransferase to dissociate to the small form of the enzyme which is the catalytically active species. Purine nucleotides cause the small active form of the enzyme to revert to the large catalytically inactive form of amidotransferase. Since the reaction catalyzed by amidotransferase

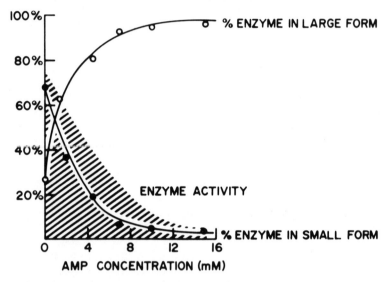

Fig. 6. Correlation between enzyme activity and the small form
of human PP-ribose-P amidotransferase. Catalytic activity was
determined at 37° during a 10 min. incubation with 2.5 mM
PP-ribose-P, 5 mM $MgCl_2$, 4 mM glutamine, 16 mM β-mercaptoethanol
and 0,2,4,7,10 and 15 mM AMP. Activity is expressed in the cross-
hatched area as a percent of maximal activity observed with 10 mM
PP-ribose-P in the absence of AMP. Aliquots of the same enzyme
preparation were incubated at 37° for 10 min. with 2.5 mM PP-ribose-P
in the presence of 0,2,4,7,10 and 15 mM AMP. These samples were
then applied to sucrose gradients that contained 2.5 mM PP-ribose-P
and the corresponding concentration of AMP. The percent of amidotransferase present as the small and large form was determined in
each gradient (small form●—●; large form ○—○). (From Holmes,
Wyngaarden and Kelley, in press).

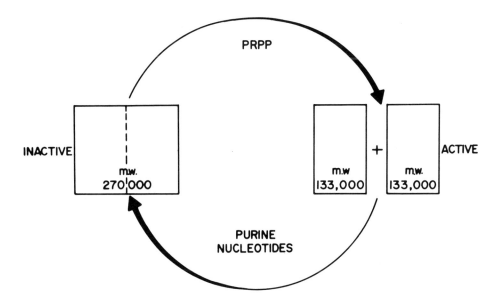

Fig. 7.   Schematic representation of the mechanism by which human
PP-ribose-P amidotransferase activity is controlled.

is the rate limiting step in the purine biosynthetic pathway, this
model provides an explanation at the molecular level for the regulation of purine biosynthesis de novo by PP-ribose-P and purine
ribonucleotides in man.

REFERENCES

Caskey, C.T., Ashton, D.M. and Wyngaarden, J.B.   1964.   The enzymol-
      ogy of feedback inhibition of glutamine phosphoribosylpyrophos-
      phate amidotransferase by purine ribonucleotides.   J. Biol.
      Chem.   239: 2570-2579.

Fox, I.H., Wyngaarden, J.B. and Kelley, W.N.   1970.   Depletion of
      erythrocyte phosphoribosylpyrophosphate in man: A newly
      observed effect of allopurinol.   New Eng. J. Med.   283: 1177-
      1182.

Fox, I.H. and Kelley, W.N.   1971.   Phosphoribosylpyrophosphate in
      man: Biochemical and clinical significance.   Ann. Intern. Med.
      74: 424-433.

Hartman, S.C. 1963. Phosphoribosylpyrophosphate amidotransferase: Purification and general catalytic properties. J. Biol. Chem. 238: 3024-3035.

Hill, D.L. and Bennett, L.L., Jr. 1969. Purification and properties of 5-phosphoribosylpyrophosphate amidotransferase from adenocarcinoma, 755 cells. Biochem. 8: 122-130.

Holmes, E.W., McDonald, J.A., McCord, J.M., Wyngaarden, J.B. and Kelley, W.N. 1973. Human glutamine phosphoribosylpyrophosphate amidotransferase: Kinetic and regulatory properties. J. Biol. Chem. 248: 144-150.

Holmes, E.W., Wyngaarden, J.B. and Kelley, W.N. Human glutamine phosphoribosylpyrophosphate amidotransferase: Two molecular forms interconvertible by purine ribonucleotides and phosphoribosylpyrophosphate. J. Biol. Chem. (in press),

Kelley, W.N., Fox, I.H. and Wyngaarden, J.B. 1970. Essential role purine biosynthesis in cultured human cells. I. Effects of orotic acid. Biochim. Biophys. Acta. 215: 512-516.

Kelley, W.N., Greene, M.L., Fox, I.H., Rosenbloom, F.M., Levy, R.I. and Seegmiller, J.E. 1970. Effects of orotic acid on purine and lipoprotein metabolism in man. Metabolism. 19: 1025-1035.

Momose, H., Nishikawa, H. and Katsujja, N. 1965. Genetic and biochemical studies of 5' nucleotide formation. II. Repression of enzyme formation in purine nucleotide biosynthesis in Bacillus subtilis and derivation of depressed mutants. J. Gen. Appl. Microbiol. 11: 211-220.

Nagy, M. 1970. Regulation of the biosynthesis of purine nucleotides in Schizosaccharomyces pombe. I. Properties of the phosphoribosylpyrophosphate glutamine amidotransferase of the wild strain and of a mutant desensitized towards feedback modifers. Biochim. Biophys. Acta. 198: 471-481.

Nierlich, D.P. and Magasanik, B. 1965. Regulation of purine ribonucleotide synthesis by end product inhibition: The effect of adenine and guanine ribonucleotides on the 5'-phosphoribosylpyrophosphate amidotransferase in Aerobacter aerogenes. J. Biol. Chem. 240: 358-365.

Reem, G.H. 1972. De novo purine biosynthesis by two pathways in Burkitt lymphoma cells and in human spleen. J. Clin. Invest. 51: 1058-1062.

Rosenbloom, F.M., Henderson, J.F., Caldwell, I.C., Kelley, W.N. and Seegmiller, J.E. 1968. Biochemical bases of accelerated purine biosynthesis de novo in human fibroblasts lacking hypoxanthine-guanine phosphoribosyltransferase. J. Biol. Chem. 243: 1166-1173.

Rottman, F. and Guarino, A.J. 1964. The inhibition of phosphoribosylpyrophosphate amidotransferase activity by Cordecepin monophosphate. Biochim. Biophys. Acta. 89: 465-472.

Rowe, P.B. and Wyngaarden, J.B. 1968. Glutamine phosphoribosylpyrophosphate amidotransferase. Purification, substructure aminoacid composition and absorption spectra. J. Biol. Chem. 243: 6373-6383.

Rowe, P.B., Coleman, M.D. and Wyngaarden, J.B. 1970. Glutamine phosphoribosylpyrophosphate amidotransferase. Catalytic and conformational heterogeneity of the pigeon liver enzyme. Biochem. 9: 1498-1505.

Schulman, J.D., Greene, M.L., Fujimoto, W.Y. and Seegmiller, J.E. 1971. Adenine therapy for the Lesch-Nyhan syndrome. Ped. Res. 5: 77.

Wood, A.W. and Seegmiller, J.E. 1973. Properties of 5-phosphoribosyl-1-pyrophosphate amidotransferase from human lymphoblasts. J. Biol. Chem. 248: 139-143.

Wyngaarden, J.B. and Ashton, D.M. 1959. The regulation of activity of phosphoribosylpyrophosphate amidotransferase by purine ribonucleotides: A potential feedback control of purine biosynthesis. J. Biol. Chem. 234: 1492-1496.

Wyngaarden, J.B. and Kelley, W.N. 1972. Gout, in The Metabolic Basis of Inherited Diseases. 3rd ed. J. B. Stanbury, J.B. Wyngaarden, and D.S. Fredrickson. (eds.). pp. 889-968. McGraw-Hill Book Co., Inc. New York.

Xanthine Oxidase

XANTHINE OXIDASE AND ALDEHYDE OXIDASE

IN PURINE AND PURINE ANALOGUE METABOLISM

Thomas A. Krenitsky

Wellcome Research Laboratories

Research Triangle Park, North Carolina   USA   27709

INTRODUCTION

Both xanthine oxidase and aldehyde oxidase catalyze the oxidation of clinically useful purines,. pteridines, and their analogues in man. In most cases, oxidation leads to inactivation of the drug [1-5].

A comparison of xanthine oxidase and aldehyde oxidase was suggested by their functional and structural similarities. Both enzymes catalyze hydroxylation reactions in which water is the source of the hydroxyl group [6,7], both have particle weights around 300,000 [8,9], and both contain FAD, molybdenum, and iron in their internal electron transport chains [7,9]. This report summarizes the results of a comparison of the distributions, substrate specificities and electron acceptor specificities of these two enzymes [10,11] and discusses the possible implications of the findings.

DISTRIBUTION

Aldehyde oxidase is not exclusively a mammalian enzyme. It was detected in many non-mammalian vertebrates and in the primitive invertebrate, the sea anemone (*Sagartia luciae*) [11]. Aldehyde oxidase, like xanthine oxidase, therefore, appears rather widely distributed in the animal kingdom. Such wide distribution suggests that the primary metabolic function of aldehyde oxidase is of a fundamental nature rather than a highly specialized one.

Among the mammals, both xanthine oxidase and aldehyde oxidase were usually most concentrated in the liver and small intestine.

57

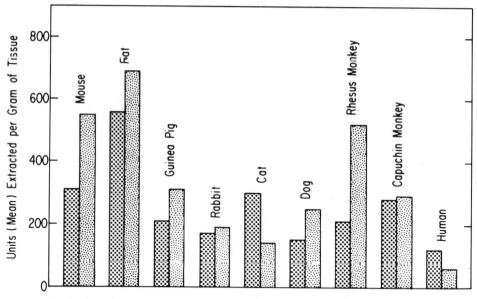

Fig. I. Xanthine Oxidase Levels in Mammalian Tissues. ⬚ Liver ⬚ Small Intestine

Figure 1[†] shows that there were appreciable differences in the levels
of xanthine oxidase in these tissues from different mammals. The
levels of activity in human liver and small intestine were low
relative to those of the other mammals studied.

The levels of aldehyde oxidase exhibited more pronounced species
differences than did those of xanthine oxidase (Figure 2). The
levels in the human, dog and cat were low as compared with the
other mammals studied, while the levels in the guinea pig and
rabbit were very high. These large species variations should be
carefully considered in metabolic studies of compounds that are
substrates for aldehyde oxidase.

SUBSTRATE SPECIFICITIES

A comparison of the substrate specificities of bovine milk
xanthine oxidase and rabbit liver aldehyde oxidase revealed both
similarities and differences [10]. Both enzymes exhibited a
preference for heterocycles which contain a condensed-pyrimidine
ring system (Figure 3)*. Of the unsubstituted ring systems studied,

---

[†]Units = 1 nanomole of product formed/minute.
*In all Figures, a wavy line at the top of the bar indicates that
the activity was below the detectable limit under the conditions
used.

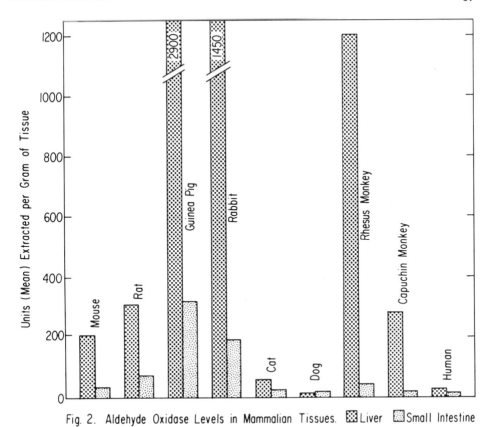

Fig. 2. Aldehyde Oxidase Levels in Mammalian Tissues. ▨Liver ▨Small Intestine

purine and pteridine were most rapidly oxidized by both enzymes.

The most distinct difference between the substrate specificities of these two enzymes was the effect of the number of C-substituents. Both enzymes readily oxidized a variety of unsubstituted and C-monosubstituted condensed-pyrimidines, but only xanthine oxidase readily oxidized any C-disubstituted derivatives. Some examples are given in Figure 4. With aldehyde oxidase, a second C-substituent always markedly decreased substrate activity, while with xanthine oxidase the substrate activity was usually relatively unaffected and, in some cases, actually increased by this substitution. The converse was found with N-substituents. Substrate activity with xanthine oxidase was usually obliterated by N-substitution, while with aldehyde oxidase many N-substituents enhanced substrate activity . As an example, the effects of N-methylation on the substrate activity of hypoxanthine with both enzymes are compared in Figure 5.

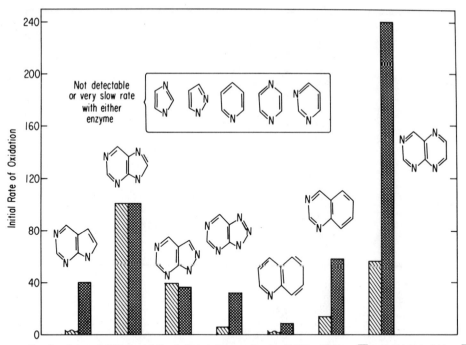

Fig. 3. Rates of Oxidation of Unsubstituted Heterocycles by Xanthine Oxidase ▧ and Aldehyde Oxidase ▨

In many cases, the chemical nature of ring substituents had markedly different effects with each enzyme. Figure 6 illustrates this point by comparing the rates of oxidation by xanthine oxidase and aldehyde oxidase of some purines substituted with different chemical groups at the 6-position.

From these specificity trends, some predictions can be made about the rate at which xanthine oxidase or aldehyde oxidase might oxidize any given heterocycle. For example, it is clear that with most C-disubstituted condensed-pyrimidines, oxidation by aldehyde oxidase in vivo is improbable, while oxidation by xanthine oxidase is a parameter which must be carefully considered.

ELECTRON ACCEPTOR SPECIFICITIES

Aldehyde oxidase in extracts of a wide variety of animal tissues did not usually use NAD^+ as an electron acceptor [11]. In contrast, with xanthine oxidase, NAD^+ was usually a relatively efficient electron acceptor. The various electron acceptor specificity patterns observed with xanthine oxidase using NAD^+, ferricyanide

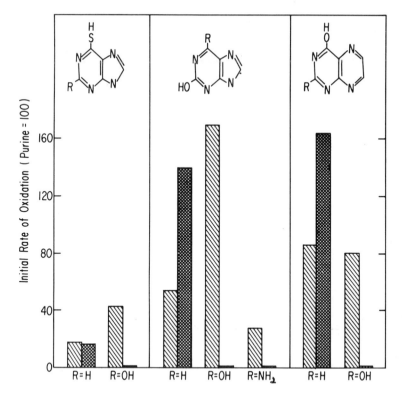

Fig. 4. Effects of a Second C-Substituent on the Rates of Oxidation of Heterocycles by Xanthine Oxidase ▨ and Aldehyde Oxidase ▦

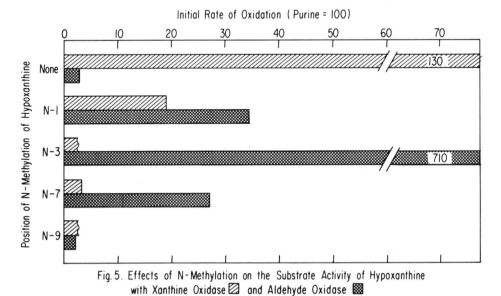

Fig. 5. Effects of N-Methylation on the Substrate Activity of Hypoxanthine with Xanthine Oxidase ▨ and Aldehyde Oxidase ▦

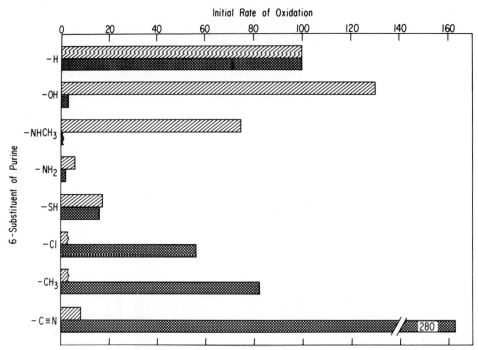

Fig. 6. Effects of the Chemical Nature of Substituents on the Rates of Oxidation of 6-Substituted Purines by Xanthine Oxidase ▨ and Aldehyde Oxidase ▩

Table 1

Electron Acceptor Specificites of Vertebrate Xanthine Oxidases

	I $NAD^+ \rangle Fe(CN)_6^{\equiv} \rangle O_2$	II $Fe(CN)_6^{\equiv} \rangle NAD^+ \rangle O_2$	III $Fe(CN)_6^{\equiv} \rangle O_2 \rangle NAD^+$
Bony fishes	+	−	−
Amphibians	+	+	−
Reptiles	+	+	−
Birds	+	−	−
Mammals	−	+	+

and oxygen are shown in Table I. Pattern I (NAD^+>ferricyanide>O_2) was found with bony fish, amphibian, reptile and bird tissue extracts. Pattern II (ferricyanide>NAD^+>O_2) was common among mammals but was also observed with a few amphibians and reptiles. Pattern III (ferricyanide>O_2>NAD^+) was only observed with some mammals, including man.

With every mammalian extract studied, the xanthine-oxidizing activities with oxygen and with NAD^+ as electron acceptors were additive. The sum of these separately measured activities closely approximated the rate of urate formation when both acceptors were present simultaneously. This additivity indicated that NAD^+ did not compete with oxygen as an electron acceptor.

Among the mammals, the relative efficiencies of NAD^+ and oxygen as electron acceptors for xanthine oxidase varied not only from species to species but also from tissue to tissue. A possible explanation of this tissue variation was suggested by the observation that aging rat liver homogenates resulted in an increase in the activity with oxygen and a decrease in the activity with NAD^+ [12]. To test the generality of this phenomenon, a variety of mammalian tissue extracts were incubated at 37°. In most cases, the result was a loss of the activity with NAD^+ and a slower loss or even increase in the activity with oxygen. This treatment, therefore, tended to convert extracts with a pattern II specificity to a pattern III specificity. The rate at which these changes occurred at 37° varied markedly from tissue to tissue. Other investigators have presented evidence indicating that changes in the electron acceptor specificity of xanthine oxidase are accompanied by some proteolysis and/or sulfhydryl modification [13,14].

COMMENTS

The physiological significance of the changes in electron acceptor specificity of xanthine oxidase in mammalian tissue extracts is uncertain at present. It is also unknown whether appreciable differences in substrate specificity accompany the changes in electron acceptor specificity. Nevertheless, the available substrate specificity data provide some basis for predicting the susceptibility of any proposed purine analogue to oxidation by either xanthine oxidase or aldehyde oxidase. Furthermore, the distribution data give some gross guidelines for predicting species differences in purine analogue metabolism. The accumulation of such knowledge is of obvious importance to the rational design and preclinical study of purine analogues of potential chemo-therapeutic usefulness.

REFERENCES

1. G. B. Elion, S. Callahan, H. Nathan, S. Bieber, R. W. Rundles,
    and G. H. Hitchings (1963) Biochem. Pharmacol., 12:85.

2. A. H. Chalmers, P. R. Knight, and M. R. Atkinson (1969) Aust. J.
    exp. Biol. med. Sci., 47:263.

3. T. L. Loo and R. H. Adamson (1962) Biochem. Pharmacol., 11:170.

4. D. G. Johns, A. T. Iannotti, A. C. Sartorelli, B. A. Booth, and
    J. R. Bertino (1965) Biochem. Biophys. Acta, 105:380.

5. M. R. Sheen, H. F. Martin, and R. E. Parks, Jr. (1970) Mol.
    Pharmacol., 6:255.

6. K. N. Murray, J. G. Watson, and S. Chaykin (1966) J. Biol.
    Chem., 241:4798.

7. K. V. Rajagopalan, I. Fridovich, and P. Handler (1962) J. Biol.
    Chem., 237:922.

8. C. A. Nelson and P. Handler (1968) J. Biol. Chem., 243:5368.

9. R. C. Bray (1963) in "The Enzymes", (P. D. Boyer, H. Lardy,
    and K. Myrbäck, eds.) 2nd ed., Vol. VII, p. 533, Academic
    Press, N. Y.

10. T. A. Krenitsky, S. M. Neil, G. B. Elion, and G. H. Hitchings
    (1972) Arch. Biochem. Biophys., 150:585.

11. T. A. Krenitsky, J. V. Tuttle, E. L. Cattau, and P. Wang
    (manuscript in preparation).

12. F. Stirpe and E. Della Corte (1969) J. Biol. Chem., 244:3855.

13. E. Della Corte and F. Stirpe (1972) Biochem. J., 126:739.

14. G. Battelli, E. Lorenzoni, and F. Stirpe (1973) Biochem. J.,
    131:191.

REGULATION OF XANTHINE DEHYDROGENASE AND PURINE NUCLEOSIDE PHOSPHORYLASE LEVELS IN CHICK LIVER

J.R. Fisher, W.D. Woodward, P.C. Lee and J. Wu

Dept. of Chemistry and Inst. of Molec. Biophys.

Florida State Univ., Tallahassee, Fla., USA

A number of agents which have been shown to affect the levels of xanthine dehydrogenase (XDH) and purine nucleoside phosphorylase (PNP) (EC 2.4.2.1) in the liver of the chick are listed in Table I. We have divided these agents into four groups depending upon

TABLE I

Effects of various agents on levels of XDH and PNP in chick liver.

Treatment		Enzyme Response[1]			References
		XDH	PNP	Other	
A	Starvation	↑ (2X)	↑ (2X)		1,2
	High Protein	↑ (6X)	↑ (3X)	↑GPA[2]	3,4,5
B	Allopurinol	↑ (3X)	→		4,6,7,8
	Adenine	↑ (2X)	→		1,4,8
C	Cortisone	↑ (2X)		↑tyrosine transaminase	9,10
	Testosterone	↓ (0.5X)			10
D	Unsaturated Fatty Acids	↓ (0.4X)	↓ (0.5X)		4,11

[1] Specific activity
[2] Glutamine - PRPP - amidotransferase

which enzymes are affected coordinately. Group A
agents induce increases in the levels of XDH, PNP and
probably other enzymes involved in purine synthesis.
So far Group B agents appear to specifically act on
XDH, although studies of other enzymes may reveal additional effects. Group C contains representative
steroid hormones which induce or suppress levels of
XDH and in one case (hydrocortisone) induces an increase
in tyrosine transaminase. Group D contains unsaturated
fatty acids including oleic, linoleic and linolenic
acids. These agents suppress levels of XDH and PNP.

MECHANISMS OF REGULATION

Group A

High protein diets have long been known to affect
the levels of various enzymes in both mammals and
birds. However, no specific mechanism of action has
as yet been suggested. We began studying this effect
using XDH in the chick and assuming that some specific
product derived from the dietary protein was causing
the level of active enzyme to increase. Feeding high
levels of several amino acids induced some significant
changes, and even ammonia caused significant increases
in some experiments. This approach did not lead to the
identification of a specific agent but suggested that
some central process in nitrogen metabolism was involved. Attempts were made to block a number of such
metabolic processes using metabolic antagonists, and it
was found that feeding transaminase inhibitors could
induce increases in XDH levels (Figure 1). In a synthetic diet containing 10% casein, aminooxy acetate induced a 3X increase in total XDH activity. Furthermore
these effects could be completely blocked using keto
acids (Table II). The levels of activity attained are
equal to the highest specific activities obtained with
dietary protein, but reductions in liver size result in
lower total activities.

On the basis of these results we are assuming that
a low molecular weight nitrogenous compound accumulates

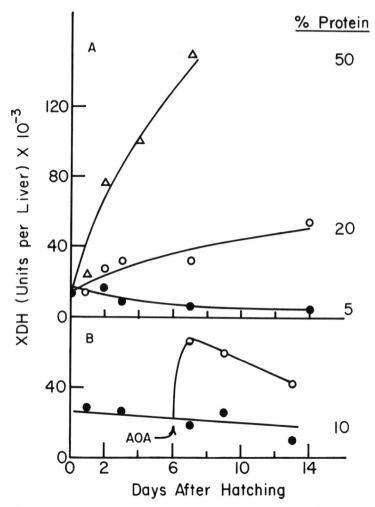

Figure 1. <u>A</u>. Effect of dietary protein on XDH accumulation in chick liver. A synthetic diet was used (15). <u>B</u>. The effect of aminooxy acetate (AOA) on XDH accumulation.

TABLE II

AOA induction of liver XDH and reversal by α-ketoglutarate.
Newly hatched chicks were starved 2 days, fed for 4
days on a synthetic diet (15) (10% casein) alone, supplemented with AOA (1 g/400 gs feed) or supplemented
with AOA and α-ketoglutarate (15 gs/400 gs feed).

Diet	XDH activity in p moles/minute	
	per gm	per organ ($\times 10^{-3}$)
Control	23.3	35.6
+ AOA	125	74.9
+ AOA and α-ketoglutarate	23.5	35.2

in the presence of transaminase inhibitors and that
this agent is an intra-cellular inducer of XDH activity.

Group B

Lee and Fisher (6) showed that feeding chicks
allopurinol induced a 3-fold increase in the level of
liver XDH (after removal of the inhibitor from the
enzyme). The rate of enzyme synthesis was elevated and
the rate of degradation unchanged. Subsequently,
DeLapp and Fisher (7) showed marked elevation of the
liver concentrations of hypoxanthine plus xanthine
(H + X) and suggested that the mechanism of induction
could involve substrate induction of enzyme synthesis.
Stirpe and Della Corte (1) have shown that adenine induces an increase in the level of liver XDH. Woodward
et al. (8) confirmed this observation and showed that
the level of H + X was also elevated. It was shown
that when adenine and allopurinol were fed together a
6-fold increase in activity could be obtained, with the
tissue levels of H + X elevated to approximately 40 mgs
per 100 gs tissue (allopurinol alone gave values of
15-20 mg % and controls were generally 1-4 mg %). It

was shown that the H + X level increased before the level of enzyme increased, which supported the substrate induction hypothesis. These results are also consistent with induction by a substrate precursor, although Weir (12) found no indication that tissue concentrations of inosine plus xanosine and their phosphate derivatives increased with allopurinol feeding. So far only one experiment seems inconsistent with the substrate induction hypothesis (12). When hypoxanthine is fed with allopurinol, it raises the hypoxanthine plus xanthine level without giving more enzyme activity (in contrast to adenine). Hypoxanthine alone does not affect either H + X or enzyme levels, presumably because it is converted to uric acid and excreted prior to reaching the liver.

At the present time we feel that the allopurinol induction effect is due to substrate induction and that the adenine effect is at least partially due to this mechanism. Proof will require in vitro studies of cells or slices.

Group C

This is a group of hormones, mostly steroid in nature, that have not yet been extensively studied in regard to their effects on liver XDH. Results with one (hydrocortisone) are shown in Figure 2. As can be seen, total activity is increased about 3-fold (yet specific activity is only doubled since this hormone causes an increase in liver size). Preliminary results indicate that levels of PNP are unaffected, which distinguishes these effects from those in Group A. Since cortisone is known to cause increased levels of liver tyrosine transaminase (9), coordinate control of XDH and amino acid degrading enzymes is a possibility.

Testosterone seems to cause a one-week delay in the developmental increase in liver XDH levels occurring after hatching. In older chicks no effect is observed. It should be noted that these studies are being made exclusively on cockerels.

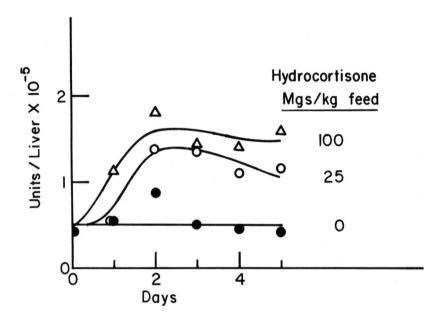

Figure 2. Effect of hydrocortisone on the levels
of liver XDH in the chick. Chicks were fed FRM starter
diet (16% protein) for 15 days, then switched to the
starter supplemented with hydrocortisone as shown.
Livers from three chicks were pooled for each assay.

Group D

Unsaturated fatty acids have been shown to suppress
XDH levels and its rate of synthesis in the chick pancreas (13,14). No effect was observed in chick liver.
However, this observation was in error because of the
age of the chicks used. When newly hatched chicks are
maintained on diets supplemented with fatty acids for
more than one week or when older chicks are used, liver
XDH levels are reduced about 50% (11). These substances
reduce the induction effect of all other agents tested,
including allopurinol, adenine, and high protein diets
(see Figures 3 and 4). In each case the fatty acid
reduced activity by about 50%.

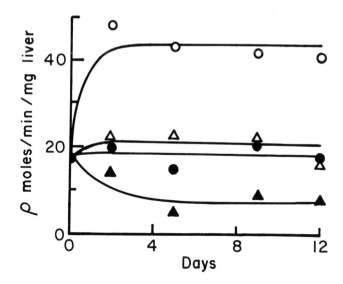

Figure 3. Effect of oleic acid (10%) on the allopurinol (0.75 gs/kg) induction of XDH in chick liver.
Chicks were fed Startena (18% protein) for three weeks
before materials were added to this diet. Controls (●),
allopurinol (O), oleic acid (▲), and allopurinol plus
oleic acid (△). The allopurinol was removed prior to
measuring XDH activities (6).

The mechanism by which these agents act is as yet
completely unknown, but it has been observed that the
pancreas concentration of the fatty acid being fed
(measured after lipid extraction and hydrolysis) is
sharply elevated and presumably the same in the liver.

SOME GENERAL COMMENTS

Regulation of XDH levels in the liver seems to be
associated with processes and conditions wherein protein
is metabolized. For example, starvation or hormonal
and dietary regimens which enhance protein catabolism
also seem to elevate XDH levels. This seems reasonable

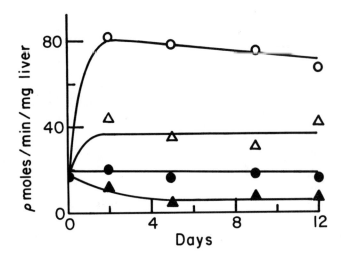

Figure 4. Effect of oleic acid (10%) on the high protein (62%) induction of XDH in chick liver. Chicks were fed Startena (18% protein) for three weeks before materials were added to this diet. Controls (●), casein (O), oleic acid (▲), and high protein plus oleic acid (Δ).

in view of the increased need to eliminate nitrogen under these conditions and the important role uric acid plays in this process in the chick. The same logic should apply to the pancreas, yet in this organ it is dietary carbohydrate which induces XDH and not protein (15,16). On high protein diets XDH levels are markedly reduced. It appears that in this organ, which is intimately involved in regulation of carbohydrate metabolism, XDH is also associated with carbohydrates in some as yet unidentified way.

XDH is an interesting enzyme to study as a model for regulation in higher organisms because it is not essential for the life of the organism (17) and therefore can be manipulated extensively, because a larger number of different agents are apparently involved in its regulation, and because different regulatory mechanisms exist in different tissues. The very preliminary results presented suggest that in liver mechanisms exist for regulating XDH independently of other enzymes (allopurinol effect), coordinately with other enzymes involved in purine metabolism (high dietary protein effect), and coordinately with enzymes involved in other aspects of nitrogen metabolism (hydrocortisone effect). So far scattered results suggest that control is at the level of enzyme synthesis. However, amino acid incorporation studies have not yet been made with the high dietary protein or hydrocortisone effects and the effects of fatty acids have only been examined properly in the pancreas (14).

ACKNOWLEDGEMENTS

This work was supported by grant HD-05544 from the U.S. Public Health Service.

REFERENCES

1. Stirpe, F. and Della Corte, E. (1965). Biochem. J. 94, 309.

2. Lee, P. C. and Fisher, J. R. (1971). Biochim. Biophys. Acta 237(1), 14.

3. Scholz, R. W. and Featherston, W. R. (1969). J. Nutr. 98, 193.

4. Lee, P. C., Nickels, J. S. and Fisher, J. R. (unpublished results).

5. Katunuma, N., Matsuda, Y. and Kuroda, Y. (1970). Adv. Enzyme Regulation 8, 73.

6. Lee, P. C. and Fisher, J. R. (1972). Arch. Biochem. Biophys. 148, 277.

7. DeLapp, N. W. and Fisher, J. R. (1972). Biochim. Biophys. Acta 269, 505.

8. Woodward, W. D., Lee, P. C., DeLapp, N. W. and
        Fisher, J. R. (1972). Arch. Biochem. Biophys.
        153, 537.

9. Chan, S. K. and Cohen, P. P. (1964).   Arch.
        Biochem. Biophys. 104, 335.

10. Wu, J. M. H. and Fisher, J. R. (unpublished
        results).

11. Wu, J. M. H. (1972). Master's Thesis, Florida
        State University, Tallahassee, Florida, USA.

12. Weir, E. E. (1972). Doctoral Dissertation, Florida
        State University, Tallahassee, Florida, USA.

13. Woodward, W. D. and Fisher, J. R. (1971). Dev.
        Biol. 24, 54.

14. Lee, P. C. and Fisher, J. R. (1971). Arch. Biochem.
        Biophys. 144(1), 443.

15. Whitney, Eleanor N., Woodward, William D. and
        Fisher, James R. (1972). J. Nutr. 102, 1347.

16. Whitney, E. N. (1973). Fed. Proc. 32, 935 abs.

17. Weir, E. and Fisher, J. R. (1970). Biochim.
        Biophys. Acta 222, 556.

XANTHINE OXIDASE ACTIVITY OF MAMMALIAN CELLS

H. Brunschede and R. S. Krooth

Columbia University, New York, N.Y.

Other authors have shown that in man xanthine oxidase is restricted to but a few tissues and is absent from lines of cultured human cells. We were able to confirm these observations and also the earlier reports that xanthine oxidase is widely distributed amongst the tissues of rodents. In addition, we have recently found that, in contrast to human cells, most lines of cultured rodent cells have high levels of xanthine oxidase activity.

Rodents are known to possess the enzyme uricase and are therefore able to carry purine catabolism one step further than man. Preliminary results suggest that uricase is restricted to but a few rodent tissues and is absent from lines of cultured rodent cells. Hence, it may be that in each vertebrate species only the final enzyme of purine catabolism is tissue restricted; it is feasible to test this generalization further. In any case, the presence of xanthine oxidase and the absence of uricase in lines of cultured rodent cells may make them useful in testing, by microbiological methods, theories about cellular factors which influence the rate of uric acid synthesis.

In other experiments, lines of mammalian cells
from several species, including man and rodents, were
grown in the presence of compounds known to induce
xanthine oxidase in a eucaryotic fungus (Aspergilus
nidulans). Uric acid and hypoxanthine were the main
inducers tested, and both were shown to enter the
cultured cells. However, neither of these compounds
led to an increase in the xanthine oxidase activity
of the mammalian cells.

These results are reported more extensively in
the following paper: Brunschede, H. and Krooth, R. S.
(1973). "Studies on the xanthine oxidase activity of
mammalian cells." Biochem. Genetics (in press).

PRPP Synthetase and PRPP

HUMAN PHOSPHORIBOSYLPYROPHOSPHATE (PP-RIBOSE-P) SYNTHETASE:

PROPERTIES AND REGULATION

I. H. Fox and W. N. Kelley

University of Toronto, Canada and Duke University

Medical Center, Durham, North Carolina 27710

PP-ribose-P and the enzyme catalyzing its synthesis, PP-ribose-P synthetase, were initially discovered by Kornberg, Lieberman and Simms in 1955. PP-ribose-P was subsequently recognized to be an essential substrate for purine, pyridine and pyrimidine biosynthesis, as well as for the formation of other compounds. Recent studies on the control of PP-ribose-P synthesis in bacteria, Ehrlich tumor ascites cells and human erythrocytes have suggested a complicated system of control depending upon ribose-5-P availability, cell energy levels and feedback inhibition from end products. Additional observations have suggested that the intracellular concentration of PP-ribose-P has a critical regulatory role in the biosynthesis of purines de novo in man. In order to clarify the regulation of PP-ribose-P synthesis, we have purified and studied this enzyme from human erythrocytes.

PP-ribose-P is synthesized from ATP and ribose-5-P (ATP: D-ribose-5-phosphate pyrophosphotransferase EC 2.7.6.1). We have assayed PP-ribose-P synthetase by two radiochemical methods (Fox and Kelley, 1971): 1) The production of PP-ribose-P and, 2) The ribose-5-P dependent conversion of ATP-^{14}C to AMP-^{14}C. Figure 1 shows that both assays are linear with time and are virtually identical using a partially purified enzyme preparation. The small discrepancy in the 2 assays is due to a loss of PP-ribose-P during the heat step in the PP-ribose-P production assay.

PP-ribose-P synthetase, purified 5100 times from human erythrocytes has been characterized (Table 1). The pH optimum was 7.2 to 7.4 and the isoelectric point was 4.7.   The enzyme exhibits an absolute requirement for inorganic phosphate, reacts equally with ATP and dATP and ribulose-5-P was 34% as

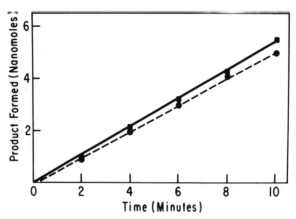

Fig. 1. Comparison of the AMP production assay and PP-ribose-P production assay for PP-ribose-P synthetase (From Fox and Kelley, 1971).

TABLE 1

CHARACTERISTICS OF HUMAN PP-RIBOSE-P SYNTHETASE

1. pH optimum - 7.2 to 7.4
2. Isoelectric point - 4.7±0.2
3. Substrate specificity:
   a. Nucleoside polyphosphates - ATP and deoxyATP
   b. Phosphorylated sugar - ribose-5-P; ribulose-5-P 34% effective
   c. Absolute requirement for P_i.
   d. $MgCl_2$ best divalent cation.
4. Km for ATP - 14 μM
        Ribose-5-P - 33 μM
        Mg - 0.2 mM
5. Distribution in man: All tissues.

effective a substrate as ribose-5-P. $MgCl_2$ was the best divalent cation.   The Km for MgATP is 14 μM, for ribose-5-P 33 μM and for Mg 0.2 mM.   PP-ribose-P synthetase is distributed in all tissues in man.

We have elucidated some of the regulatory properties of human PP-ribose-P synthetase by studying the reversible association and dissociation of the enzyme, the kinetic mechanism, and end product inhibition (Fox and Kelley, 1971; Fox and Kelley, 1972).

Multiple molecular forms of PP-ribose-P synthetase have been studied by gel filtration for Stokes radius determinations, by sucrose gradient ultracentrifugation to measure the sedimentation coefficients, and by sodium dodecyl sulfate polyacrylamide disc gel electrophoresis to quantitate the basic subunit molecular weight (Table 2). The aggregated form of the enzyme occurs in the presence of 0.3 mM ATP and 6 mM Mg. Using Stokes radius values and sedimentation coefficients of the native forms of PP-ribose-P synthetase, a range of apparent molecular weights from 60,000 to 1,200,000 can be estimated. Subunit molecular weight was estimated to be 34,500 in sodium dodecyl sulfate. Apparent molecular weights observed in the presence of other denaturing solvents, sodium chloride and urea, gave intermediate values.

Several additional conclusions are possible based on other

TABLE 2

APPARENT MOLECULAR WEIGHTS AND FRICTIONAL RATIOS OF SOME OF THE
FORMS OF PP-RIBOSE-P SYNTHETASE

Form	Stokes Radius	$S_{20,w}$	Molecular Weight	f/f_0
	A	$X\ 10^{13}$ sec		
A. Native forms				
1. Phosphate and EDTA	29	5.0	60,000	1.13
2. ATP-Mg				
Peak 1	116.5	24.3	1,162,000	1.68
Peak 2	99	17.7	720,000	1.68
B. Denaturing solvents				
1. Sodium chloride	31.7	4.1	53,000	1.28
2. Urea	30	3.5	43,000	1.30
3. Sodium dodecyl sulfate[a]			34,500	

[a] Evaluated by acrylamide gel electrophoresis
    (From Fox and Kelley, 1971)

observations: a) the associated form of the enzyme appears to be
the active form, b) the associated enzyme appears to be a more stable
form   at temperatures ranging from -70°C to 60°C, c) PP-ribose-P
synthetase is probably aggregated in human erythrocytes since
concentrations of ATP and MgCl$_2$ which aggregate the enzyme approximate the range of normal values.

    The kinetic mechanisms of PP-ribose-P synthetase partially
purified from human erythrocytes has been studied. Double reciprocal plots of initial velocity studies with variable MgATP cencen-
trations and fixed ribose-5-P are intersecting and suggest a
sequential kinetic mechanism, (Figure 2). Product inhibition studies

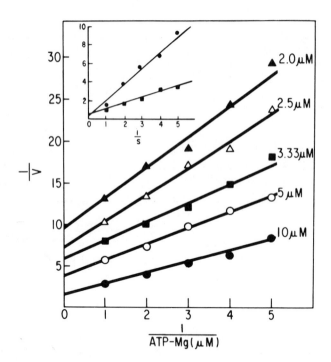

Fig. 2.   Double reciprocal plot of initial velocity studies with
variable MgATP concentrations and fixed ribose-5-P concentrations
ranging from 2 to 10 μM. Magnesium concentration is 6 mM.   Inset
shows a secondary plot of the slope (■-■) and the intercept
(●-●) versus the inverse of ribose-5-P concentration in micromolar.
The Km for MgATP is 14 μM and ribose-5-P is 33 μM (From Fox and
Kelley, 1972).

have shown that inhibition by AMP is non-competitive with respect to both substrates, while PP-ribose-P inhibits in a manner which is competitive with respect to ribose-5-P and non-competitive with respect to MgATP. These observations are most compatible with an ordered BiBi mechanism in which ribose-5-P binds first and PP-ribose-P is released last. The reaction mechanism suggested by our data is:

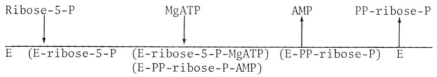

Initial velocity studies with magnesium complicate this postulated mechanism by implicating a separate sequential binding of this divalent cation to PP-ribose-P synthetase, in addition to its well known complex formation with ATP.

    Human PP-ribose-P synthetase is inhibited by a large number of end products of purine, pyrimidine, and pyridine metabolism. In general terms, the di-and triphosphate derivatives were more potent inhibitors than the monophosphates. If inhibitor concentration is progressively elevated to a high unphysiological concentration, inhibition continues to increase until in some cases almost complete inhibition occurs (Table 3). This virtually

TABLE 3

EFFECT OF SOME INHIBITORS AT SATURATING SUBSTRATE CONCENTRATIONS[a]

Concentration[b]	Percentage Inhibition		
	1 mM	5 mM	10 mM
PP-ribose-P	19	49	65
AMP	53	73	86
ADP	92	98.5	100
GMP	3	23	35
GDP	24	89	100
IMP	0	5	11
IDP	24	98	
ITP		81	100
CMP		5	7
CDP		23	40
NMN	1	19	
NADH	7	35	65

[a]MgATP, 0.3 mM, ribose-5-P, 0.5 mM.
[b]Magnesium concentration adjusted for each concentration of inhibitor to give a concentration of magnesium which is 6 mM in excess of total starting nucleotide concentration. (From Fox and Kelley, 1972).

eliminates the possibility of a cumulative mechanism of inhibition
since this mechanism requires that maximal inhibition by any one
end product is substantially less than 100%. The compounds
studied in pairs produced inhibition which was equal to or less
than the total inhibition produced by both studied independently,
but greater than either alone (Fig. 3). In no case was the
inhibition by a pair of inhibitors greater than the sum of the
inhibition produced by each alone, a finding incompatible with
a synergistic mechanism of inhibition. These inhibitors appear
to interact with PP-ribose-P synthetase at one of three different
sites (Table 4). ADP inhibits PP-ribose-P synthetase in a manner
competitive with respect to MgATP and the Ki for ADP (0.01 mM) is
below its intracellular concentration at least in the erythrocyte.
This may reflect regulation of the enzyme by cellular energy
levels as previously suggested by Atkinson and Fall (1967).

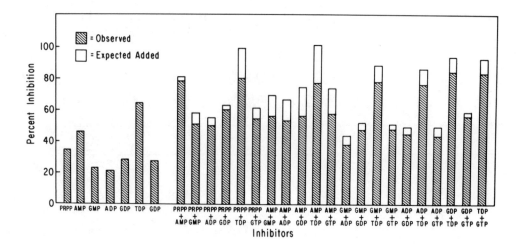

Fig. 3. Effect of inhibitors of PP-ribose-P synthetase alone or in
pairs. Inhibitors of PP-ribose-P synthetase were studied in the following concentrations: PP-ribose-P (PRPP), 0.05 mM; AMP, 0.5 mM;
GMP, 1.0 mM, ADP, 0.005 mM; GDP, 0.2 mM; TDP, 0.1 mM; GTP, 0.25 mM.
Inhibition of the enzyme was studied with each inhibitor alone or
paired with another. In the latter case the total inhibition (cross-
hatched bars) was compared to the sum of the inhibition caused by
each inhibitor alone (clear bars). The observed inhibition by paired
inhibitors was always less than the sum of both inhibitors, but
greater than each alone (From Fox and Kelley, 1972).

TABLE 4

MECHANISM OF INHIBITION OF PP-RIBOSE-P SYNTHETASE

Inhibitor	Variable substrate	K_i Slope	Type of Inhibition
		mM	
Products			
PP-ribose-P	MgATP	0.10	Noncompetitive
	Ribose-5-P	0.05	Competitive
AMP	MgATP	0.2	Noncompetitive
	Ribose-5-P	0.4	Noncompetitive
End Products			
ADP	MgATP	0.01	Competitive
	Ribose-5-P	0.07	Noncompetitive
NAD	MgATP	0.7	Noncompetitive
GDP	MgATP	0.1	Noncompetitive
	Ribose-5-P	0.3	Noncompetitive
GTP	MgATP	0.2	Noncompetitive
GMP	MgATP	2.0	Noncompetitive
2,3-DPG	Ribose-5-P	5.3	Competitive

(From Fox and Kelley, 1972).

PP-ribose-P and 2,3-DPG both inhibit the enzyme by a mechanism which is competitive with respect to ribose-5-P. This may provide a link between oxygenation of hemoglobin and regulation of PP-ribose-P synthesis in the erythrocyte. Most compounds studied were found to inhibit the enzyme non-competitively by a third mechanism which corresponds to heterogeneous nucleotide pool inhibition. This type of inhibition has been found for PP-ribose-P synthetase from Salmonella typhimurium and implies that this group of inhibitors bind at one site on the enzyme and have a low affinity for this site (Switzer, 1967).

The regulation of PP-ribose-P synthetase may be important in the control of specific pathways for which PP-ribose-P is a substrate. Both PP-ribose-P amidotransferase and PP-ribose-P synthetase are inhibited by the same end products and under certain conditions some of these products such as AMP, ADP, GDP, and GTP are more potent inhibitors of PP-ribose-P synthetase. It is also clear that the concentration of PP-ribose-P plays a crucial role in regulating PP-ribose-P amidotransferase and the entire purine biosynthetic pathway. These observations suggest that the activity of PP-ribose-P synthetase may be important in the regulation of human purine biosynthesis de novo.

REFERENCES

Atkinson, D. E. and Fall, L. 1967. Adenine triphosphate conserva-
    tion in biosynthetic regulation: Escherichia coli phosphori-
    bosylpyrophosphate synthetase. J. Biol. Chem. 242: 3241-3242.

Fox, I. H. and Kelley, W. N. 1971. Human phosphoribosylpyrophos-
    phate synthetase: Distribution, purification and properties.
    J. Biol. Chem. 246: 5739-5745.

Fox, I. H. and Kelley, W. N. 1972. Human phosphoribosylpyrophos-
    phate synthetase: Kinetic mechanisms and end-product inhibi-
    tion. J. Biol. Chem. 247: 2126-2131.

Kornberg, A., Lieberman, I. and Simms, E. S. 1955. Enzymatic
    synthesis and properties of 5-phosphoribosylpyrophosphate.
    J. Biol. Chem. 215: 389.

Switzer, R. L. 1967. End product inhibition of phosphoribosyl-
    pyrophosphate synthetase. (Abstract). Fed. Proc. 26: 560.

REGULATORY ASPECTS OF THE SYNTHESIS OF 5-PHOSPHORIBOSYL-1-PYROPHOSPHATE IN HUMAN RED BLOOD CELLS.

Nira Sciaky, A. Razin, Bruria Gazit and J. Mager

Department of Cellular Biochemistry

The Hebrew University-Hadassah Medical School, Jerusalem

It has been established in earlier investigations[1,2] that the synthesis of purine nucleotides in mammalian red blood cells (RBC) is governed by the extent of endogenous supply of 5-phosphoribosyl-1-pyrophosphate (PRPP), as an essential intermediary. Consequently, the elucidation of the mechanisms controlling the formation of PRPP within the cell appears to be of crucial importance for the understanding of the overall metabolic regulation of purine nucleotide biosynthesis.

Previous data[1] indicated that the rate of synthesis of PRPP from glucose in RBC incubated in isologous plasma or in physiological saline solution was negligibly low, unless the suspending medium was fortified with unphysiologically high amounts of inorganic phosphate (Pi). The concentration of Pi required for optimal PRPP formation in the intact cells exceeded by a factor of tenfold the normal blood Pi level and was nearly 4 times higher than the corresponding value for PRPP synthetase, as measured in a cell-free system. However, the intracellular synthesis of PRPP at the optimum level of Pi amounted to about 0.5% only of the rate observed in equivalent amounts of stroma-free hemolysate supplemented with ribose-5-phosphate (Rib-5-P) (fig.1.)

The PRPP yield in the RBC was strikingly enhanced by addition of methylene blue or phenazine methosulfate. These artificial hydrogen acceptors stimulate the oxidative pentose cycle pathway and thereby increase the intracellular supply of Rib-5-P as the requisite substrate of PRPP synthetase. However, the stimulatory effect of the redox dyes failed to alter the stringent requirement for high Pi concentration. On the other hand, experiments performed with cell-free systems revealed that the activity of PRPP synthetase

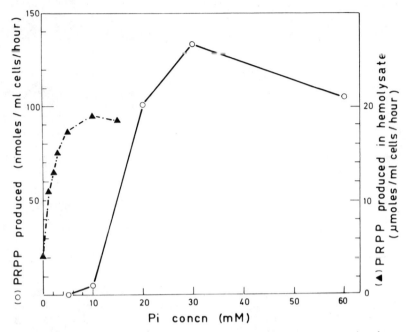

Fig. 1. Effect of Pi concentration on PRPP formation in intact
erythrocytes and in stroma-free hemolysates. Packed washed human
erythrocytes were suspended in an equal volume of a mixture consis-
ting of 0.9% NaCl solution and 0.1M potassium phosphate buffer (pH
7.4) in suitable proportions to give the various concentrations of
Pi, as indicated. The cell suspensions were incubated in the presence of 10mM glucose for 60 min at 37°C with continous shaking.
Intracellular PRPP and PRPP synthetase in the cell-free hemolysates
were assayed as previously described[1].

was severely inhibited by ADP, AMP and 2,3-diphosphoglycerate and
that this inhibition was relieved by high concentrations of Pi[1].

    The overall data, therefore, strongly favored the conclusion
that the intracellular output of PRPP is primarily controlled by
the allosteric devices regulating the activity of PRPP synthetase,
rather than by the availability of Rib-5-P. It seemed also reasonable to postulate the existence of some alternate, physiological
mechanism for overcoming the regulatory restraint and enabling PRPP
generation to proceed at a pace commensurate with the rate of its
consumption in the process of nucleotide turnover[3,4].

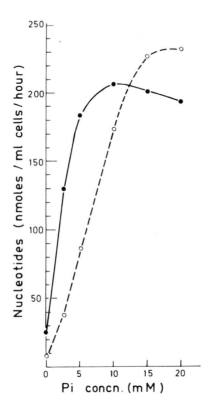

Fig. 2. Effect of anaerobiosis on the Pi requirement for PRPP formation in intact RBC. Experimental conditions were as described in the legend to Fig. 1. O-------O incubation in air; ●————● incubation under nitrogen.

In view of the known binding capacity of deoxygenated hemoglobin for 2,3-diphosphoglyceric acid (2,3-DPG) and adenine nucleotides[5],[6] it appeared likely that the inhibitory action of these
effectors and accordingly the Pi requirement for the synthesis of
PRPP in RBC would be considerably diminished under anaerobic conditions. This prediction was borne out by the data presented in Fig.
2, which showed that formation of PRPP at low Pi levels was indeed
much more efficient under an atmosphere of nitrogen than in air.

In an attempt to gain further insight into the regulatory
mechanisms operative within the cell, we decided to study the synthesis of PRPP from glucose in a cell-free system imitating the
conditions prevailing in the intact RBC.

In this system (for details see legend to Table 1) about 10-15%
of the ATP added was found to be split to ADP and AMP during the 60
min incubation period employed in the experiment. Under these conditions, the formation of PRPP at 2mM Pi and in the absence of an ATP
regenerating system, was reduced by the products of ATP breakdown
(ADP and AMP) to less than 10% of the activity obtained in the
presence of the ATP regenerating system. This inhibition was largely abolished at 20mM Pi. It should be also noted that the inhibition was substantially attenuated when the Mg^{2+} concentration was
increased from 2mM (close to the physiological level) to 5mM
(Table 1).

Addition of 2,3-DPG to the same system in the amount of 5mM
(which is well within the physiological range of this substance in
the RBC) resulted in complete obliteration of PRPP synthesis, when
tested at 2mM Pi and in the absence of the ATP regenerating system.
The superimposed inhibition of 2,3-DPG was not appreciably allevia-
ted by raising the Pi concentration to 20mM (Table 2).

Essentially similar results were obtained, when glucose was
replaced by Rib-5-P. The parallel behavior of the two substrates
seemed, therefore, to support the notion that the site of inhibition
was located at the stage of PRPP synthetase. It may be pertinent to
refer in this context to a closely analogous control mechanism operating at the level of hexokinase which has been described by Brewer[6].

Using the same reaction mixture as above but omitting NADP, so
as to eliminate the oxidative pentose cycle pathway, we were able
to reproduce essentially the characteristic response to Pi observed
under anaerobic conditions in the intact RBC. As may be seen in
Table 3, there was a definite shift of the Pi optimum towards physiological concentrations. The inhibitory effect of 15mM Pi seems
to be attributable to interference of the Pi excess with the activity of the transaldolase-transketolase system[7] mediating the non-
oxidative pentose cycle path for Rib-5-P supply.

Table 1. Synthesis of PRPP from glucose or Rib-5-P in stroma-free hemolysates in the presence and in the absence of ATP-regenerating system; effect of Pi and Mg^{2+} concentrations.

| Substrate | ATP regenerating system | PRPP produced (μmoles/ml RBC) | | | |
| | | 2mM Mg^{2+} | | 5mM Mg^{2+} | |
		2mM Pi	20mM Pi	2mM Pi	20mM Pi
Glucose	+	1.80	1.70	1.70	1.65
Glucose	−	0.12	1.41	0.65	1.40
Rib-5-P	+	4.80	4.60	5.20	5.10
Rib-5-P	−	0.25	3.60	1.52	5.10

The reaction mixture contained in a final volume of 0.2 ml the following components: 50mM Tris (pH7.4), 50mM KCl, 1mM ATP, 2mM GSH, 0.2mM NAD, 0.05mM NADP, 2.5mM glucose or 0.5mM ribose-5-phosphate (as specified), 0.5mM 8^{14}C-hypoxanthine (0.5μc/μmole), and 40μl of 1:4(vol/vol) charcoal-treated and dialysed human red-cell hemolysate. The ATP-regenerating system consisted of 2.5mM creatine phosphate and 50μg/ml creatine kinase. The reaction mixture was incubated for 60 min at 37°. The nucleotide produced was isolated on a plastic sheet coated with PEI-cellulose (Machery Nagel Co.), and counted in a liquid scintillation spectrometer.

Table 2. Effect of 2,3-diphosphoglycerate (DPG) on PRPP formation in human red-cell hemolysates.

| Substrate | ATP regenerating system | PRPP produced (μmoles/ml RBC) | | | |
| | | 2mM Pi | | 20mM Pi | |
		no DPG	5mM DPG	no DPG	5mM DPG
Glucose	+	1.3	0.8	1.2	0.8
Glucose	−	0.1	0.0	1.2	0.9
Rib-5-P	+	3.6	2.4	3.2	2.5
Rib-5-P	−	0.3	0.0	2.8	2.2

Reaction mixture and other details were as described in the legend to Table 1. The concentration of Mg^{2+} was 2mM.

Table 3. Synthesis of PRPP from glucose dependent on the non-oxidative pentose cycle; effect of Pi.

Pi (mM)	Without ATP regenerating system	With ATP regenerating system
	nmoles PRPP/ml RBC/hr	
2	105	520
5	250	520
15	205	380

Conditions as in Table 2.

In trying to extrapolate the results to the conditions in vivo we suggest that the low oxygen tension prevailing in the venous circulation favors the formation of PRPP at the physiological level of Pi by diminishing the inhibitory activity of the allosteric effectors (ADP and 2,3-DPG). This interpretation is in line with the observation that RBC affected with severe glucose-6-phosphate dehydrogenase deficiency, and therefore essentially incapable of utilizing the oxidative pentose phosphate cycle for Rib-5-P formation, show no significant impairment of their PRPP-synthesizing activity from glucose[1]. It is possible, however, that some additional and as yet undetermined factor may be instrumental in vivo in relieving the allosteric inhibition of the PRPP synthetase.

REFERENCES

1. A. Hershko, A. Razin and J. Mager. Biochim. Biophys. Acta, 184, 64, (1969).
2. A. Hershko, Ch. Hershko and J. Mager. Israel J. Med. Sci., 4, 939, (1968).
3. J. Mager, A. Dvilansky, A. Razin, E. Wind and G. Izak. Israel J. Med. Sci. 2, 297, (1966).
4. J. Mager, A. Hershko, R. Zeitlin-Beck, T. Shoshani and A. Razin. Biochim. Biophys. Acta, 149, 50, (1967).
5. R. Benesch, R. E. Benesch and C. D. Yu. Proc. Nat. Acad. Sci. U. S., 59, 531, (1968).
6. G. J. Brewer. Biochim. Biophys. Acta, 192, 157, (1969).
7. Z. Dische and D. Igals. Arch. Biochem. Biophys, 101, 489, (1963).

This work was supported by a grant from the National Research Council for Research and Development and an award to one of us (J. M.) from the Chief Scientist's Office of the Ministry of Health. The excellent assistance of Miss Magda Benedikt is gratefully acknowledged.

PHARMACOLOGICAL ALTERATIONS OF INTRACELLULAR PHOSPHORIBOSYLPYROPHOSPHATE (PP-RIBOSE-P) IN HUMAN TISSUES

I. H. Fox and W. N. Kelley

University of Toronto, Canada and Duke University Medical
Center, Durham, North Carolina 27710

Intracellular PP-ribose-P, a ribose sugar with phosphate at the five position and pyrophosphate at the 1 position (Fig. 1), is complexly regulated by a balance between synthesis and degradation. Any alteration in PP-ribose-P concentration may potentially alter the rate of purine biosynthesis de novo since PP-ribose-P is rate limiting for this pathway. Drug induced changes of intracellular PP-ribose-P may in this way substantially alter human purine metabolism.

Pharmacological intervention in PP-ribose-P metabolism may occur at a number of sites (Figure 2). PP-ribose-P is synthesized from ribose-5-P and ATP. The availability of ribose-5-P plays a critical role since the intracellular concentration of this compound is well below the Km for PP-ribose-P synthetase. Alterations in the rate of PP-ribose-P synthesis could also be produced by changes in Mg, Pi, cell energy levels or the concentration of nucleotides in the cell. PP-ribose-P is utilized in two major ways. The introduction of excess substrate for the phosphoribosyltransferase enzymes could lower PP-ribose-P levels by consuming it in the formation of a ribonucleotide derivative and inhibiting PP-ribose-P synthesis by increasing the nucleotide pool. The rate of PP-ribose-P hydrolysis by alkaline and acid non-specific phosphatases may be altered by changes in Pi and nucleotide levels which normally exert an inhibitory effect.

We have studied several compounds which inhibit purine biosynthesis de novo in cultured human fibroblasts and in vivo and are known to react with PP-ribose-P in the presence of phosphoribosyltransferase (PRT) enzymes to form ribonucleotide derivatives (Fox and Kelley, 1971).

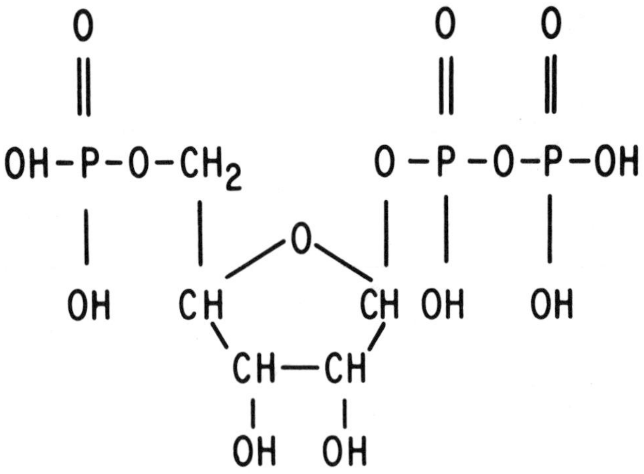

$$OH-\overset{\overset{\displaystyle O}{\|}}{P}-O-CH_2 \qquad O-\overset{\overset{\displaystyle O}{\|}}{P}-O-\overset{\overset{\displaystyle O}{\|}}{P}-OH$$

α-5-phospho-D-ribosyl-l-pyrophosphate
(PRPP)

Fig. 1.   Structure of phosphoribosylpyrophosphate (PRPP).

The administration of allopurinol at a dose ranging from 2 to 4 mg/Kg to 9 patients with normal hypoxanthine-guanine phosphoribosyltransferase (HGPRT) led to a significant reduction in intracellular PP-ribose-P levels with a maximum decrease to a mean of 47% of control values observed within three to five hours (Fig. 3). Allopurinol also depleted intracellular PP-ribose-P levels in vitro in intact erythrocytes with normal HGPRT. Additional studies have demonstrated that the depletion of PP-ribose-P produced by allopurinol is due to the conversion of allopurinol to its ribonucleotide, a process which utilizes PP-ribose-P. The depletion of intracellular PP-ribose-P and the formation of allopurinol ribonucleotide, a known inhibitor of PP-ribose-P amidotransferase, may both contribute to the decreased rate of purine biosynthesis de novo commonly observed in patients on allopurinol therapy.

Orotic acid, an intermediate of pyrimidine biosynthesis de novo, reacts with PP-ribose-P to form orotidine-5-monophosphate in

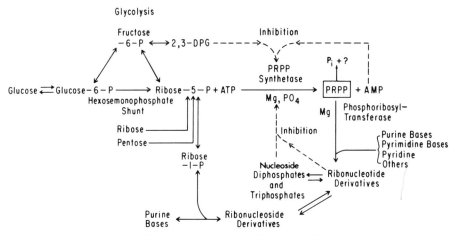

Fig. 2.   Regulation of intracellular phosphoribosylpyrophosphate (PRPP) concentration.

the presence of orotate phosphoribosyltransferase (OPRT). The administration of orotic acid 1 gram in vivo to two patients with normal PP-ribose-P levels led to a depletion of erythrocyte PP-ribose-P (Fig. 3). Purine biosynthesis de novo, as determined by following the incorporation of isotopic glycine into urinary uric acid, was inhibited during orotic acid therapy in 2 patients. Diminished PP-ribose-P levels and inhibition of purine biosynthesis have also been observed following the addition of orotic acid in cultured human fibroblasts.

Adenine and one of its analogs, 2,6-diaminopurine, react with PP-ribose-P in the presence of adenine phosphoribosyltransferase (APRT) to form their respective ribonucleotide derivatives. Both of these compounds were reported by Schulman, et al. (1971) to cause a substantial drop in erythrocyte PP-ribose-P levels when administered to a patient with the Lesch-Nyhan syndrome. In 1 gouty patient given adenine 1 gram, a maximum decrease in erythrocyte PP-ribose-P levels to 25% of control values occurred within three hours (Fig. 3). Adenine has also been shown to be an inhibitor of purine synthesis de novo in normal and gouty subjects (Seegmiller, et al., 1968).

Nicotinic acid, a pyridine compound, reacts with PP-ribose-P in the presence of nicotinate phosphoribosyltransferase (NPRT) to form nicotinic acid mononucleotide. The administration of nicotinic acid 1 gram to 5 patients was followed by a decrease in erythrocyte PP-ribose-P concentration to 25% of control values (Fig. 4). Nicotinic acid caused a quantitatively lesser reduction of PP-ribose-P in erythrocytes studied in vitro. Nicotinic acid has been shown by others to inhibit purine biosynthesis de novo in cultured fibro-

blasts (Boyle, et al., 1972) but not <u>in vivo</u> (Becker, et al., 1973).
These studies demonstrate that a number of drugs which are substrates
for a PRT enzyme may reduce intracellular PP-ribose-P and potentially decrease the rate of purine biosynthesis.

Two other compounds have been observed to diminish PP-ribose-P

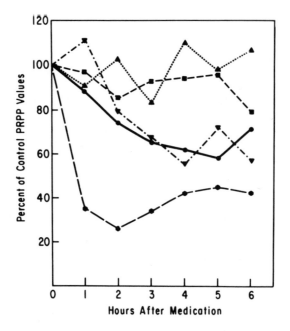

Fig. 3. Effect of administration of drugs on intracellular phosphoribosylpyrophosphate (PP-ribose-P) levels in erythrocytes from
patients with normal hypoxanthine-guanine phosphoribosyltransferase
activity. Each point represents the mean value observed in the
patients studied (expressed as percent of control values) from 1 to
6 hr after drug administration: ●—●, allopurinol (2 to 4 mg/Kg
body weight, 9 patients) orally: ▲▲ oxipurinol (4 to 8 mg/Kg body
weight, 4 patients) orally: ▼▼ orotic acid (2 g, 2 patients) orally:
■■ methylene blue (2 mg/Kg body weight, 1 patient) intravenously;
and ◆—◆ adenine (1 g, 1 patient) orally. (From Fox and Kelley, 1971)

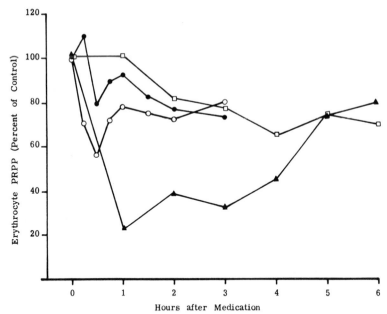

Fig. 4. Effect of administration of drugs on intracellular phosphoribosylpyrophosphate (PP-ribose-P) levels in erythrocytes from patients with normal hypoxanthine-guanine phosphoribosyltransferase activity. Each point represents the mean value observed in the patients studied (expressed as percent of control values) from 1 to 6 hours after drug administration. △—△ nicotinic acid (1 gram, 5 patients) orally, □—□ nicotinamide (1 gram, 3 patients) orally, ○—○ fructose (0.5 mg/Kg, 4 patients) intravenously, ●—● 2-deoxyglucose (50 mg/Kg, 1 patient) intravenously.

levels in vivo perhaps by a different mechanism. Intravenous fructose, 0.5 g/Kg administered to 3 patients, caused a mean 30% decrease in PP-ribose-P levels after 30 minutes (Fig. 4). Parental 2-deoxyglucose, 50 mg/Kg, reduced intracellular PP-ribose-P by 27% after three hours in 1 patient (Fig. 4). However, the exact mechanism by which these compounds reduce PP-ribose-P is unclear.

An increased production of PP-ribose-P has been produced in vitro by several compounds which increase ribose-5-P availability and could potentially accelerate the rate of purine biosynthesis de novo. Methylene blue, 10 to 100 μM, has been shown to increase intracellular PP-ribose-P in Ehrlich tumor ascites cells, in human erythrocytes, and in diploid human fibroblasts. In the latter cells a concomitant increase in purine biosynthesis de novo was observed. Intravenous methylene blue (2 mg/Kg) given to 2 patients did not alter erythrocyte PP-ribose-P levels (Fig. 3). Glucose, fructose and ribose also increase erythrocyte PP-ribose-P levels in vitro. However, as noted above, fructose infusions in vivo had an opposite effect.

PP-ribose-P has been reported by others to be elevated by nucleosides and the trophic hormones, ACTH and TSH,which also presumably act by increasing ribose-5-P availability (Fox and Kelley, 1971). In addition, Estrogen has been observed to stimulate the synthesis of PP-ribose-P synthetase in rat uterus (Oliver, 1972). Finally, inhibition of PP-ribose-P hydrolysis provides a potential means of elevating intracellular PP-ribose-P levels (Fox and Marchant, in preparation).

In the present manuscript we have summarized the evidence to suggest that drug mediated changes in purine metabolism can, in many instances, be related to alterations in the intracellular concentration of PP-ribose-P. Many of the ways in which PP-ribose-P levels may be manipulated are listed in Table 1.

TABLE 1

PHARMACOLOGICAL ALTERATIONS OF PP-RIBOSE-P AVAILABILITY IN HUMAN
TISSUES

DECREASED PP-RIBOSE-P CONCENTRATION:
  Increased Utilization:
    Adenine
    Allopurinol
    2,6-Diaminopurine
    Nicotinic Acid
    Orotic Acid
    Others - Potentially any compound that reacts with PP-ribose-P
            to form a ribonucleotide derivative
  Inhibition of Synthesis:*
    2-Deoxyglucose
    Fructose (in vivo)
  Increased Degradation:
    Divalent cations - Non-Enzymatic**

INCREASED PP-RIBOSE-P CONCENTRATION:
  Stimulation of Synthesis:
    Glucose, Fructose, Ribose
    Methylene Blue
    Trophic hormones--ACTH, TSH
    Estrogen
    Others - Potentially any compound that can increase synthesis
            of ribose-5-P.
  Inhibition of Hydrolysis:*

---

* Mechanism uncertain
** Mechanism is not established experimentally in vivo or in intact
   cells in vitro.

REFERENCES

Becker, M. A., Raivio, K. O., Meyer, L. J. and Seegmiller, J. E. 1973. Effect of nicotinic acid on human purine metabolism. Fed. Proc. 21: 616 (Abstract).

Boyle, J. A., Raivio, K. O., Becker, M. A. and Seegmiller, J. E. 1972. Effect of nicotinic acid on human fibroblast purine biosynthesis. Biochim. Biophys. Acta. 260: 179-183.

Fox, I. H. and Kelley, W. N. 1971. Phosphoribosylpyrophosphate in man: Biochemical and clinical significance. Ann. Int. Med. 74: 424-433.

Fox, I. H. and Marchant, P. J. Phosphoribosylpyrophosphate hydro-lyzing activity: Characteristics and distribution in human tissues. (in preparation).

Gershon, S and Fox, I. H. Effects of nicotinic acid on uric acid metabolism in man. Arth. Rheum. (Abstract) (in press).

Oliver, J. M. 1972. A possible role for 5-phosphoribosyl-1-pyro-phosphate in the stimulation of uterine purine nucleotide synthesis in response to oestradiol-17B. Biochem. J. 128: 771-777.

Schulman, J. D., Greene, M. L., Fujimoto, W. Y. and Seegmiller, J. E. 1971. Adenine therapy for Lesch-Nyhan syndrome. Ped. Res. 5: 77-82.

Seegmiller, J. E., Klinenberg, J. R., Miller, J. and Watts, R.W.E. 1968. Suppression of glycine-N^{15} incorporation into urinary uric acid by adenine-8-^{13}C in normal and gouty subjects. J. Clin. Invest. 47: 1193-1203.

Nucleoside and Nucleotide Metabolism

PURINE SALVAGE IN SPLEEN CELLS

Partsch,G.,[1) Altmann,H.,[1) and Eberl,R.[2)

1) Institute of Biology, Research Centre Seibersdorf, A-2444 Seibersdorf, Austria
2) II.Med.Department for Rheumatic Diseases,
   Krankenhaus Lainz, Vienna, Austria

Since the discussion livened up again about the formation of
nucleotides for DNA-replication and DNA-repair by the contested
theory of Werner (1971) it seems important to study the problems
of nucleotide formation during repair processes. Our own studies performed with the radiosensitive organism Byssochlamys
fulva and the radioresistant Pullularia pullulans led us to
suppose that the synthesis of purine nucleoside monophosphates
via purinephosphoribosyltransferase (purine-PRT) is not as
sensitive to radiation as the de novo synthesis. Therefore
a relation between the salvage pathway and nucleotide formation during repair processes does not seem impossible.

Because of the rapid increase of chemical substances and
pharmaceuticals which inhibit DNA-repair in mammalian cells,
efforts are made to understand the mechanisms of such inhibitors. The need for knowledge about DNA-repair affecting substances is apparent when refering to the human skin disease
Xeroderma pigmentosum, where malignancy often occurs as an
after effect. Among the substances found up till now to inhibit DNA-repair mechanisms, commonly used detergents in pharmaceuticals eg Tween 80 should be pointed out (Gaudin 1971,
Tuschl 1972). But there are also a great number of other substances like mycotoxins, carcinogenic carbohydrates and other
environmental pollutants under discussion. Developing a system
to screen possible DNA-repair inhibitors, human peripheral
leucocytes or mouse spleen cells have been used in our laboratory (Kocsis 1972). Subsequently studies were initiated to investigate the influence of such inhibitors on purine nucleotide formation. In the first stages these experiments were
done with mouse spleen cells and the role of purine-PRT and

other purine salvage pathways was studied.

Materials and Methods

For the experiments 3 month old female Swiss mice
(HA/ICR) were used. After taking out the spleen, the organ
was homogenized with 3 ml Tris-HCl buffer (0,1 M + 5 mM $MgCl_2$,
pH 7,4) in a Potter cooled with ice. The broken cells were
centrifuged for 30 minutes at 30.000 x g and the supernatant
used for the enzyme determinations. Protein was measured after
the method of Lowry (1951).

Purine-PRT Activity

The method used was an in vitro system containing C^{14}-
labeled purine bases. The nucleoside monophosphates formed
during the reaction were fixed on DEAE paper discs (Serva).
The reaction mixture (total of 200 μl) consisted of:
175 μl Tris-HCl buffer (see above), 5 μl PRPP (2 mg/100 μl
buffer, tetrasodium salt, Sigma) and 10 μl purine base
(= 0,25 μCi C^{14}-8-adenine (NEN Chemicals, 6,53 mCi/mM), C^{14}-8-
guanine (NEN Chemicals, 5,4 mCi/mM) or C^{14}-8-hypoxanthine
(NEN Chemicals, 4,18 mCi/mM)). After addition of all the components (in an ice bath), 10 μl of the homogenate was quickly
added and after vigorous shaking 25 μl of the reaction mixture was immediately pipetted onto a DEAE paper disc.

The remaining mixture was incubated at 37°C and further
samples were taken at 15, 30 and 45 minutes. The paper discs
were immediately washed twice in 10 mM ammoniumformiate, then
in aqua dest. and finally in 96 % ethanol. After drying, the
remaining radioactivity was estimated in a liquid scintillation counter (Packard) using POPOP (0,3 g/l) and PPO (7,0 g/l)
in toluol.

GMP-Nucleotidase

For the determination of the nucleotidase 180 μl substrate (1 ml Tris-HCl buffer + 10 μl H^3-GMP (= 1 μCi, NEN
Chemicals, 1,23 Ci/mM) were supplemented with 20 μl of the
cell homogenate. For the remaining procedure see above. The
decrease in radioactivity caused by the cleavage of the 5'-
phosphate group with time was followed by liquid-scintilla-
tion counting.

Indirect Assay for Purine Nucleoside Phosphorylase-
and Purine Deoxynucleoside Phosphorylase Activity

The determination of purine nucleotide formation from purine bases in the presence of ribose-1-phosphate and deoxyri-
bose-1-phosphate was carried out using the same reaction mixture as for the purine-PRT activities, but instead of PRPP,
5 μl ribose-1-phosphate (5 mg/100 μl, dicyclohexylammonium
salt, Sigma) or 5 μl deoxyribose-1-phosphate (2 mg/100 μl, di-
monocyclohexylammónium salt, Sigma) were used. In both cases
the purine bases guanine, hypoxantine and adenine were tested.

Results and Discussion

First of all it should be noted that because of its in-
homogeneity mouse spleen is not the ideal material for these
experiments. The spleen contains a lot of erythroblasts, mega-
caryocytes, plasma cells, lymphocytes, leucocytes and erythrocytes which of course have different activities of purine
salvage enzymes.

Table 1

The purine-PRT activities in various cells expressed
in nmol transformed base per mg protein per hour, calculated
by regression analysis and compared with results reported in
the literature.

	G-PRT	H-PRT	A-PRT
mouse spleen cells	708	276	125
human erythrocytes			
Kelley (1969)	103 ±21	103 ±18	31,1 ±6
Sperling (1970)	112,8 ±22,3	50,09±	10,44±2,4
Partsch (in preparation)	92,56± 7,7	96,86±17,3	-
human leucocytes			
Kelley (1969)	183	128	211
hen erythrocytes			
Partsch (1973)	4,74± 2,45	2,65± 2,2	15,47±3,98

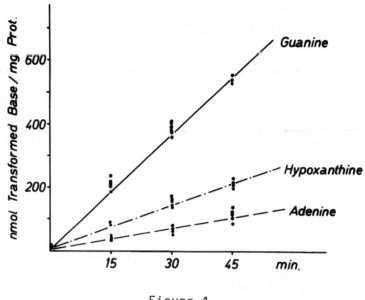

Figure 1

The purine-PRT activity in mouse spleen cell homogenates

As expressed in table 1 the purine-PRT activities show
great differences within cells of the same organism (human
erythrocytes and leucocytes) and also between cells of dif.
ferent organisms. This variability is specific to the organism
and in the case of human erythrocytes depends on the method
used. Comparing the results obtained from mouse spleen cells
with those of humans and microorganisms (Partsch 1971) one can
note that the enzyme activity in spleen cell homogenates is
much higher than in human. Guanine-PRT shows the highest activity followed by hypoxanthine-PRT and adenine-PRT respectively.

This fact indicates that spleen cells possess a highly
active salvage pathway similar to that found in microorganisms
(Partsch 1971). But in contrast microorganisms de novo synthesis may be absent, as is the case in rabbit bone marrow cells
(Thompson 1960) and human leucocytes (Scott 1962).

A strong catabolism due to purine nucleotidase is not the
reason for the high PRT activities, because the level of GMP-
nucleotidase is very low.

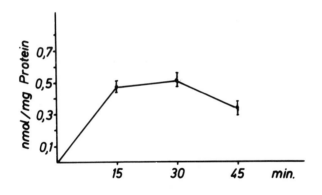

Figure 2

Formation of guanosine by 5'-nucleotidase
from H^3-GMP in mouse spleen homogenates (nmol/mg protein)

     Due to this finding addition of thymidine triphosphate
for the inhibition of nucleotidase activities in the purine-
PRT estimation was omitted. This statement is correct only
for GMP-nucleotidase but just guanine was transformed strongly
in spleen homogenate.

     Another interesting point is that in human erythrocytes
the guanine-PRT and hypoxanthine-PRT activity is similar
(Kelley 1969) or nearly similar (Partsch, in preparation) but
in mouse spleens guanine-PRT is 2,5 times more active than
hypoxanthine-PRT.

     Therefore the question arises whether two separate enzymes
are responsible for the transformation of guanine and hypoxanthine.

     Speaking about purine salvage it implies not only the pu-
rine-PRT and its pathological appearance in human, but also
includes all other possible reutilisation of nucleotide precursors derived from catabolic processes. Gallo (1971) reports
that a second pathway exists in human leucocytes.

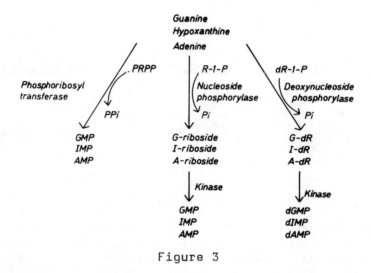

Figure 3

Purine salvage pathways existent in mouse spleen cells

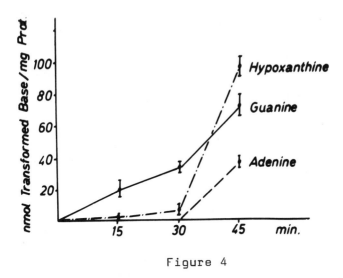

Figure 4

Nucleosidephosphorylase activity in mouse spleen cells

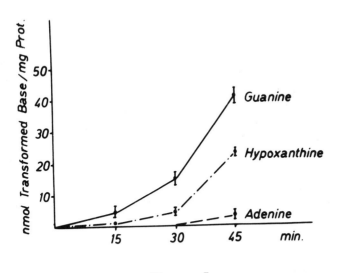

Figure 5

Deoxynucleosidephosphorylase activity in mouse spleen cells

During this pathway the purine bases react with ribose-1-phosphate or deoxyribose-1-phosphate, forming an intermediate (nucleoside or deoxynucleoside) which is further phosphorylated by kinases.

Testing this with spleen cell homogenates it could be evidenced that both reactions are possible for guanine, hypoxanthine and also adenine. In human leucocytes however Gallo (1969) found that adenine was not a substrate for deoxynucleasidephosphorylase.

As expressed in figures 4 and 5 the enzyme activities are lower than the corresponding purine-PRT activities. In both cases the formation of AMP and dAMP was measured only after 45 minutes of incubation. This perhaps is due to the induction of one of the enzymes.

Summarizing we can say that normally the low activity of nucleoside phosphorylase and deoxynucleosidephosphorylase seems to exclude their effective participation in the formation of precursors for DNA-repair processes. But for example at high radiation doses when DNA is damaged and deoxyribose units and other degredation products (deoxyribose donors) are present, the deoxyribosephosphorylase may play an important role in nucleotide formation. Further investigations with these enzymes using homegenous cell material, for example human lymphoid cells, should give more details and also provide an answer as to whether there is any co-operation of purine-PRT in repair processes.

This work was supported by the Fonds zur Förderung der wissenschaftlichen Forschung in Österreich.

Gallo,R.C. and Perry,S., 1969, J.Clin.Invest. 48, 105.

Gallo,R.C., 1971, Acta haemat. 45, 136.

Gaudin,D., Gregg,R.S. and Yielding,K.L., 1971, Biochem. Biophys.Res.Comm. 45, 630.

Kelley,W.N., Greene,M.L., Rosenbloom,R.M., Henderson,J.F. and Seegmiller,J.E. 1969, Annals Int.Med. 70, 155.

Tuschl,H., Klein,W., Kocsis,F., Bernat,E. und Altmann,H.: SGAE-Bericht Nr.2107, BL-59/73, März 1973.

Kocsis,F., Klein,W. und Altmann,H.: SGAE-Bericht Nr.2055, BL 50/72, November 1972.

Lowry,O.H., Rosenbrough,N.J., Farr,A.L. and Randall,R.J., 1951. J.Biol.Chem. 193, 265.

Partsch,G., Altmann, H., 1973, Experentia 29:267

Partsch,G., Altmann, H., 1971, IAEA Symposium Wien, 53.

Partsch,G., Frank,O. and Altmann,H., in preparation

Scott,J.L., 1962, J.Clin.Invest. 41, 67.

Sperling,O., Frank,M., Ophir,R., Liberman,U.A., Adam,A. and deVries,A. 1970. Europ.J.Clin.Biol.Res. 15, 942.

Thomson,R.Y., Ricceri,G., Peretta,M. 1960. Biochem.Biophys. Acta 45, 87.

Werner, Nature New Biol. 233. 99-103, 1971.

PURINE NUCLEOTIDE SYNTHESIS, INTERCONVERSION AND CATABOLISM IN

HUMAN LEUKOCYTES

J. F. Henderson, E. E. McCoy and J. H. Fraser

Cancer Research Unit and Department of Pediatrics

University of Alberta, Edmonton, Alberta, Canada

Purine metabolism in man is of increasing clinical and experimental interest, but studies of this subject have been hampered by methodological difficulties. This paper describes new methods for the measurement of apparent activities of more than 20 enzymes of purine metabolism in human leukocytes. These methods have also been applied to cultured human lymphoblasts and monolayer cultures of skin fibroblasts and amnionic cells.

Mixed leukocytes are prepared using dextran sedimentation of erythrocytes, centrifugation of leukocytes from the plasma, and lysis of residual erythrocytes with ammonium chloride. Granulocytes and mononuclear leukocytes are separated from each other and from erythrocytes by density gradient centrifugation using a mixture of Ficoll, Hypaque and dextran. Thirty ml of blood provides sufficient leukocytes for study of adenine, guanine and hypoxanthine metabolism in duplicate. Suspension cultures of human lymphoblasts and leukemic cells, and monolayer cultures of skin fibroblasts and amnionic cells were studied under normal culture conditions; approximately 10^6 cells are required for each incubation.

Leukocytes from blood are suspended (4% by volume) in 0.1 ml of Fischer's medium containing 25 mM phosphate buffer, pH 7.4, and incubated for 20 minutes. Radioactive adenine, guanine or hypoxanthine (50 µM final concentration) are added to separate tubes and incubation continued for 3 hours. For suspension cultures, 10^6 cells are suspended in 2.5 ml of fresh growth medium, cells are incubated 20 minutes, a radioactive purine is added (20 µM final concentration), and incubation continued for 30 minutes. For monolayer cultures, a radioactive purine (20 µM final concentration) is added to a non-confluent culture containing about

113

10^6 cells and 3 ml of Eagles' medium, and incubation continued for 60 minutes.

Radioactive metabolites are extracted with perchloric acid, and extracts are neutralized with potassium hydroxide. Purine bases and ribonucleosides are separated by thin-layer chromatography on cellulose, and purine ribonucleotides are separated by thin-layer chromatography on PEI-cellulose. Radioactivity in each compound is measured.

The pathways of purine metabolism studied are shown in the following diagram:

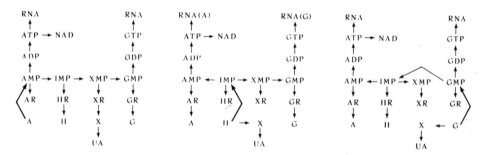

Adenine metabolism   Hypoxanthine metabolism   Guanine metabolism

The apparent activity of each enzyme of purine ribonucleotide metabolism is determined by calculating the amount of radioactive substrate metabolized by each enzyme; i.e., radioactivity in all metabolites further along a pathway from the enzyme in question is summed. These calculations are performed by computer. As an example, the sums used when radioactive hypoxanthine is substrate are as follows:

Enzyme	Sum
H PHOSPHORIBOSYL-TRANSFERASE	IMP+AMP+ADP+ATP+NAD+XMP+GMP+GDP+A+AR+HR+X+XR+G+GR+N.A.
AMPS SYNTH. + LYASE	AMP+ADP+ATP+NAD+N.A. −A+A+AR
AMP KINASE	ADP+ATP+NAD+N.A.−A
ADP KINASE	ATP+NAD+N.A.−A
IMP DEHYDROGENASE	XMP+GMP+GDP+GTP+XR+G+GR+N.A.−G
GMP SYNTHETASE	GMP+GDP+GTP+G+GR+N.A.−G
GMP KINASE	GDP+GTP+N.A.−G
GDP KINASE	GTP+N.A.−G
RNA POLYMERASE	N.A.−A+N.A.−G

AMP DEPHOSPHORYLASE	A+AR
IMP DEPHOSPHORYLASE	HR
XMP DEPHOSPHORYLASE	XR
GMP DEPHOSPHORYLASE	G+GR
AR PHOSPHORYLASE	A
GR PHOSPHORYLASE	G
XANTHINE OXIDASE	X+UA

These methods permit the measurement of the apparent activities of more than 20 enzymes of purine metabolism. The cell preparations used are relatively pure, the metabolic integrity is assured by the high ATP/ADP ratios which are measured. However, the assumptions and limitations of these procedures must be recognized. For example, it is assumed that all pathways are unidirectional and linear, that all reactions and metabolites are measured, and that the system is homogeneous. The rates of individual reactions may not be linear over the period studied, total radioactivity in each metabolite is measured rather than total amount or specific activity, and the statistical error is different for each reaction. Finally, nucleotide synthesis from purine bases is a limiting step. The magnitude of these limitations can be estimated and taken into account in interpreting the data obtained.

Normal values for enzymes of adenine, guanine and hypoxanthine metabolism have been determined for mixed blood leukocytes from 30 individuals. The normal range of values is calculated as ± 2 standard deviations from the mean. As an example, the results obtained when hypoxanthine is used as precursor are shown in Table 1.

Replicate analyses done the same day vary by less than 10%. Analyses of leukocytes from single individuals taken on different days reveal a greater variation, the basis of which is under study. For example, values for adenine phosphoribosyltransferase activity in mixed leukocytes of a single individual, measured on different days were 4056, 5980, 3742, 8334, 7922 and 5963 nmoles per 10^{10} cells.

These procedures were validated by study of leukocytes and erythrocytes of a patient with the Lesch-Nyhan syndrome. Age, sex, time and temperature of blood storage (up to 3 hours, 4° or room temperature) do not appear to affect these values. Other potential variables are under investigation.

Measurement of apparent activities of enzymes of purine metabolism have been made in mixed blood leukocytes from more than 100 patients with severe mental retardation of unknown etiology, but in which a hereditary element is possible. About 10% showed

TABLE 1 - HYPOXANTHINE METABOLISM IN LEUKOCYTES

Reaction	Normal Values (nmoles/10^{10} cells)	
	Mean	Range
H phosphoribosyltransferase	1661	810-2500
AMPS synth. + lyase	695	333-1100
AMP kinase	565	233-900
ADP kinase	497	210-780
IMP dehydrogenase	402	80-720
GMP synthetase	358	50-670
GMP kinase	95	8-190
GDP kinase	59	21-97
AMP dephosphorylase	100	0-220
IMP dephosphorylase	389	110-660
XMP dephosphorylase	34	0-74
GMP dephosphorylase	248	0-520
AR phosphorylase	29	0-59
GR phosphorylase	98	0-290
X oxidase	161	0-450

one or more enzyme activity outside the normal range and about
half of these have had consistently elevated values in multiple
experiments. For example, values for adenine phosphoribosyltransferase in mixed leukocytes of one patient, measured on different
dates, were 15334, 9314, 10680 and 11595 nmoles per 10^{10} cells; all
of these values are well above the normal range. No patients have
been detected with consistently lowered values of any enzyme. In
at least some cases, elevated purine phosphoribosyltransferase
activities in intact cells are accompanied by increased enzyme
activity in cell extracts. Unfortunately, normal concentrations
of phosphoribosyl pyrophosphate are too low to be measured with the
methods presently available.

Application of these methods to cultured human lymphoblasts
have detected individual differences in their purine metabolism.
Initial experiments using monolayer cultures have shown that these
methods are also applicable to such systems and do not require dis-
lodgement of the cells from the flask. Hence they should be
applicable to the study of purine metabolism in cells cultured from
skin biopsies and amniocentesis samples.

INCORPORATION OF PURINE ANALOGS INTO THE NUCLEOTIDE POOLS OF

HUMAN ERYTHROCYTES

R. E. Parks, Jr., Phyllis R. Brown and Chong M. Kong

Division of Biological and Medical Sciences
Brown University, Providence, R. I. 02912, U.S.A.

INTRODUCTION

For the past decade this laboratory has devoted much of its attention to an examination of various facets of purine metabolism in human erythrocytes. These cells do not have the complete pathway for the de novo synthesis of purines and do not make nucleic acids. On the other hand, they have an active nucleotide metabolism and contain the salvage enzymes, hypoxanthine-guanine phosphoribosyl transferase (HGPRTase), adenine phosphoribosyl transferase (APRTase) and adenosine kinase. In view of the fact that the activities of certain enzymes of purine metabolism are quite high (e.g., purine nucleoside phosphorylase occurs at a level of about 15 umolar units/ml of erythrocytes) and the total mass of erythrocytes in the adult human being is in excess of two liters, it appears that these cells play an important and perhaps not yet fully appreciated role in the whole body economy of purines in man. Therefore, we believe that the human erythrocyte provides a very useful model system for the examination of purine metabolism in man as well as for investigations of the action of certain purine and purine nucleoside antimetabolites, many of which are important in medicine.

The present report summarizes some of our recent investigations into the incorporation of natural purine bases and nucleosides, as well as related analogs, into the nucleotide pools of human erythrocytes. A number of these studies were greatly facilitated by the recent development of high pressure liquid chromatography which makes it possible to examine intracellular

nucleotide profiles rapidly and on small amounts of tissue
(Brown, 1970; Scholar et al., 1973; Brown et al., 1972; Parks et
al., 1973). Through these studies, human erythrocytes have been
shown capable of incorporating surprisingly large amounts of
certain nucleosides into the nucleotide pools. Also included are
the results of an examination of the effect of pH on the erythrocytic HGPRTase reaction with guanine and several purine base analogs.

STUDIES OF THE EFFECT OF pH ON THE ERYTHROCYTIC HYPOXANTHINE-
    GUANINE PHOSPHORIBOSYL TRANSFERASE (HGPRTase) REACTION

    Although a number of reports published earlier of studies
with various preparations of HGPRTase have indicated that the pH
of the reaction medium may influence the reactivity of the enzyme
with various purine analogs (Way and Parks, 1958; Miller and
Bieber, 1968; Krenitsky et al., 1969), there have been no publi-
cations describing the effects of pH on the kinetic parameters
of the reaction. Therefore, a Dixon pH analysis (Dixon, 1952)
has been undertaken to examine the effect of pH on the maximal
velocities (V_{max}) and Michaelis constants (K_m) of a partially
purified preparation of human erythrocytic HGPRTase with guanine
and the purine analogs, 6-thioguanine, 6-selenoguanine, 6-mer-
captopurine and 8-azaguanine as substrates. Since these compounds have strikingly different acid dissociation constants
(pK_a's), i.e., 8-azaguanine, 6.5; 6-mercaptopurine, 7.7; 6-seleno-
guanine, 7.6; 6-thioguanine, 8.2; and guanine, 9.2, a study of
this type offers an opportunity to examine the effect of the
ionic state of the purine ring on the enzymatic reaction. In
order to perform this experiment, specific spectrophotometric
assays were developed for each of the substrates and the specific
change in the molar absorptivity at selected wavelengths and at
each pH during the conversion of the base to the 5'-monophosphate
ribonucleotide was determined. Figure 1 presents the results of
a Dixon analysis in which the logarithms of the maximal velocities
and the negative logarithms of the Michaelis constants (pK_m's)
were plotted at various pH values. No striking differences were
observed in V_{max} values with the various substrates when studied
at a particular pH value within the pH range from 6.5 to 9.3.
However, as pH increased over this range, a corresponding increase in V_{max} values was observed (Figure 1a). Examination at
pH values greater than 9.3 was not possible due to inactivation
of the enzyme. This observation of increasing V_{max} with increasing pH is in agreement with the finding of Miller and Bieber
(1969) who observed marked differences in the activation energy
of the reaction of yeast HGPRTase at different pH values. The
plot of pK_m vs. pH (Figure 1b) yielded a family of graphs with
downward curvatures. It is of considerable interest that with

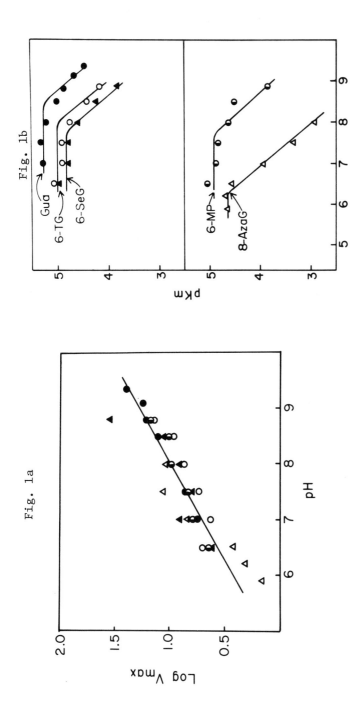

FIGURE 1. Dixon analysis of the reactivity of human erythrocytic HGPRTase. Using specific spectrophotometric assays, the K_m and V_{max} values were determined at the indicated pH's with a partially purified preparation of human erythrocytic HGPRTase. The variable substrates were: 8-azaguanine (8-azaG); 6-mercaptopurine (6-MP); 6-selenoguanine (6-SeG); 6-thioguanine (6-TG); guanine (Gua). The details of this work will be published elsewhere.

the exception of guanine, when the breakpoint of the graph for
each substrate was determined by the method of Dixon, it occurred
at a pH value close to the pK_a of the particular substrate. In
this type of analysis a downward curvature indicates that the
ionization of this group is either in the substrate or in the free
enzyme. The fact that different breakpoints were obtained with
the same enzyme but with different substrates strongly indicates
that the ionizing group is located in the substrate and that the
active form of the substrate is the unionized form. This conclusion was supported by the finding that, when the data were
recalculated on the basis of the quantity of unionized substrate present at each pH value, a family of reasonably straight
lines was obtained. The fact that the curve for guanine has a
downward break not associated with its pK_a value suggests the
possibility that ionization of the free enzyme may also play a
role at higher pH's. However, this possibility was difficult
to study because of instability of the enzyme in the region of
pH 9 and above. A complete description of these studies is in
preparation and will be submitted for publication (Kong and Parks,
unpublished results).

INCORPORATION OF NATURAL PURINE RIBONUCLEOSIDES AND THEIR ANALOGS
        INTO THE NUCLEOTIDE POOLS OF HUMAN ERYTHROCYTES

    1. Incorporation of Guanosine, Inosine and their Analogs

    The recent development of high pressure liquid chromatography
(HPLC) has made possible the rapid and reproducible examination
of many of the components of the free nucleotide pools on quantities of tissues as small as 1 mg (Horvath et al., 1967; Brown,
1970; Brown and Parks, 1973). In comparison with most other tissues examined in this laboratory, the nucleotide profiles of
freshly drawn and washed human erythrocytes are relatively un-
complicated and consist predominantly of the adenine nucleotides
and little or no guanine nucleotides. However, human erythrocytes contain the complete enzymatic pathway for the conversion
of guanosine or guanine to GTP (Agarwal et al., 1971). When
washed human erythrocytes are incubated for two hours in the
presence of relatively high concentrations of guanosine and inorganic phosphate (1.0 millimolar and 20-30 millimolar, respectively), the synthesis of impressive quantities of GDP and GTP
occurs. As shown in Figure 2, amounts of GTP and GDP can accumulate that exceed by several fold the concentrations of ATP
and ADP. A possibility of considerable interest to be explored
in future studies is that the accumulation of these large quantities of high-energy phosphate in the form of GTP and GDP can
improve the storage of blood under blood bank conditions.

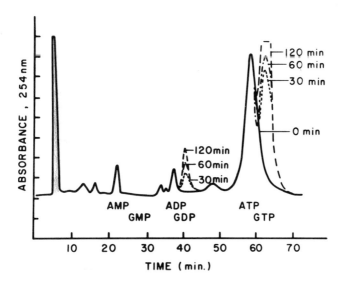

FIGURE 2.   Synthesis of guanine nucleotides from guanosine by human erythrocytes.   A 33% suspension of erythrocytes was incubated in a medium containing 128 mM NaCl, 1.2 mM MgCl$_2$, 18 mM potassium phosphate (pH 7.4), 16 mM glucose and 0.5 mM guanosine at 37° C.   At 0, 30, 60 and 120 minutes, 1 ml of the suspension was pipetted slowly into 2 ml of 12% trichloroacetic acid (TCA) with rapid stirring.   After centrifugation, 0.5 ml aliquots of the supernatant fluid were extracted 3 times with 5 ml H$_2$O-saturated diethylether to remove TCA.   The extract was analyzed by high pressure liquid chromatography as described by Brown (1970).

In contrast to findings with guanosine, when adenosine is used as precursor, there is observed the formation of large quantities of IMP as has been described elsewhere (Parks and Brown, 1973; Manohar et al., 1968; Meyskens and Williams, 1971). In studies that are still preliminary, after incubation of human erythrocytes with inosine, in addition to the formation of a large quantity of IMP, new small peaks appeared on HPLC profiles with the retention times of IDP and ITP.   If these peaks are, in fact, IDP and ITP, their formation will be consistent with the recent observation that human erythrocytic guanylate kinase, in contrast to the enzyme from other tissues, has weak substrate activity with IMP (Agarwal et al., 1971; Agarwal and Parks, 1972).

In view of the fact that a number of hypoxanthine and guanine analogs are excellent substrates for erythrocytic HGPRTase and the ribonucleosides are substrates for erythrocytic purine nucleoside phosphorylase (PNPase) (Ross et al., 1973; Parks and Agarwal, 1972), it is suspected that analog nucleotides might be formed in human erythrocytes. Earlier attempts to examine this question were hampered by the precipitation of the thiolated purines and their derivatives, presumably through the formation of covalent linkages with denatured proteins during the extraction of reaction mixtures containing human erythrocytes. To avoid this problem, incubations and extractions were performed in the presence of dithiothreitol (2 mM).

When suspensions of human erythrocytes were incubated with 6-thioguanosine and extracts were subjected to ion exchange chromatography on Dowex-1-formate columns according to the method of Moore and LePage (1958), the elution patterns of Figure 3 were obtained. Here, there is clear evidence for the formation of 6-thioGMP with the gradual production of increasing quantities of 6-thioGTP upon prolonged incubation. The quantities of nucleotides produced are approximately 0.23 micromoles of 6-thioGMP and 0.17 micromoles of 6-thioGTP per milliliter of erythrocytes at 2 hours. When similar experiments were performed with the analog nucleosides, 6-selenoguanosine and 6-thioinosine, evidence for the formation of substantial quantities of 5'-monophosphate ribonucleotides of the analogs was obtained. However, in contrast to the findings with 6-thioguanosine, prolonged incubation with 6-selenoguanosine or 6-thioinosine did not lead to the formation of the analog triphosphate nucleotides. A manuscript describing the details of these studies is in preparation (Kong and Parks, unpublished results).

2. Incorporation of Adenosine and its Analogs into Human
    Erythrocytic Nucleotide Pools

When freshly drawn and washed human erythrocytes are incubated with adenosine, HPLC profiles reveal a large new peak in the position of IMP. However, no significant change was detected in the concentrations of ADP or ATP (Parks and Brown, 1973). This finding is in accord with those of other laboratories (Manohar et al, 1968; Meyskens and Williams, 1971) and indicates that under the conditions employed most of the adenosine that enters the cell is deaminated by adenosine deaminase to form inosine which, in turn, is split by purine nucleoside phosphorylase to liberate hypoxanthine. The hypoxanthine can react with HGPRTase and PRPP to form IMP. Recent studies with a series of adenosine analogs are in accord with this suggestion. The substrate activities have been determined with a preparation of adenosine deaminase purified about 3000 fold from human erythrocytes (Agarwal et al., 1973). Among the analogs tested,

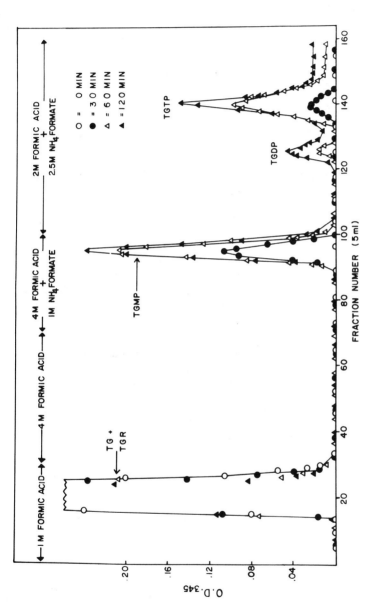

FIGURE 3. Incorporation of 6-thioguanosine into the nucleotide pools of human erythrocytes. The medium contained 75 mM NaCl, 50 mM potassium phosphate (pH 7.4), 2 mM glucose, 2 mM dithiothreitol, 1 mM 6-thioguanosine, with 25% erythrocytes. Incubation was carried out at 37° C. A 5 ml aliquot was removed at the indicated time and added slowly to 2.5 ml of 12% cold perchloric acid with rapid mixing. The extract was centrifuged and the supernatant was neutralized with KOH. After removing precipitated potassium perchlorate, the neutralized extract was analyzed by anion-exchange chromatography using the method of Moore and LePage (1958).

8-azaadenosine and formycin A were found to have V_{max} values
several times greater than that of adenosine. No formation of
nucleotides was detected with either analog compound in intact
erythrocytes. On the other hand, analogs such as tubercidin,
2-fluoroadenosine and 6-methyladenosine were not substrates for
adenosine kinase and when incubated with intact erythrocytes
entered the nucleotide pools readily and gave evidence of the
formation of substantial amounts of the triphosphate nucleotide
(Agarwal et al., 1973; Parks and Brown, 1973). Of special interest is the adenosine analog, 2-fluoroadenosine, because its
nucleotides have retention times that are about ten minutes longer than their normal adenine counterparts with the elution system employed in this laboratory (Brown, 1970). Figure 4 presents the results of HPLC of trichloroacetic acid extracts of
human erythrocytes incubated with 2-fluoroadenosine for two hours.
The quantity of F-ADP and F-ATP formed exceeds the concentration
of the normal nucleotide counterparts by several fold. In time
studies, it appears that in normal intact erythrocytes the rate

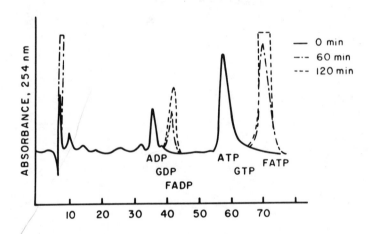

FIGURE 4. Synthesis of polyphosphate nucleotides of 2-fluoro-
adenosine in human erythrocytes. A 20% suspension of freshly
washed erythrocytes was incubated at 37° C in 10 ml of medium
containing 108 mM NaCl, 30 mM potassium phosphate buffer (pH 7.4),
10 mM glucose, and 1 mM 2-fluoroadenosine. A 1 ml sample was removed at the indicated time and added dropwise with rapid stirring
to 1 ml of cold 15% trichloroacetic acid (TCA). After freezing
and thawing twice, protein was removed by centrifugation. The
TCA in 1 ml of supernatant was removed from the solution by extraction with H_2O-saturated ether (3 x 10 ml) and 20 ul were analyzed by high pressure liquid chromatograph as described by
Brown (1970).

of 2-F-ATP synthesis is approximately 0.027 umoles per minute
per milliliter of erythrocytes. Interestingly, a similar rate of
uptake of labeled adenosine into human erythrocytes was determined by Manohar et al. (1968). In studies of the source of the
high-energy phosphate employed for the synthesis of this new nucleotide, surprisingly it was found that the concentrations of
ATP and 2,3-diphosphoglycerate remained unchanged during nucleotide synthesis. However, the rate of lactate production increased from a normal control value of about 2.4 micromoles per
hour per milliliter of cells to 5.9 umoles/hr/ml of erythrocytes.
Thus it appears that the flux of erythrocytic glycolysis can be
increased sufficiently to accommodate this new demand for high-
energy phosphate. The details of these studies with adenosine
analogs will be published (Parks and Brown, 1973).

DISCUSSION

In comparison with other tissue the nucleotide profiles of
freshly drawn normal human erythrocytes are relatively simple and
consist predominantly of adenine nucleotides with only small
amounts of nucleotides that contain guanine, uridine or cytosine.
However, the complete enzymatic pathway for the conversion of
guanine or guanosine to GTP is found in these cells (Agarwal et
al., 1971). It is now shown through these studies that human
erythrocytes can rapidly synthesize large quantities of GDP and
GTP, as well as the nucleotides of analogs of guanosine, inosine
and adenosine. These findings raise many intriguing questions
about the role of erythrocytes in whole body purine metabolism,
as well as in the chemotherapeutic and immunosuppressive action
of purine antimetabolites, some of which play a prominent role in
medicine. An intriguing hypothesis is suggested by the observation that compounds that are not substrates for adenosine deaminase often can form nucleotides, presumably by reaction with adenosine kinase and in some cases form very large amounts of the polyphosphate nucleotides (Figure 4). Recent studies indicate that
the activity of adenosine deaminase is about 0.15 umolar units
per milliliter of human erythrocytes and that the Michaelis constant for adenosine is about 1.1×10^{-4} M (Agarwal et al., 1973).
It appears that in the formation of F-ATP from 2-F-adenosine, the
enzyme adenosine kinase catalyzes the rate-limiting reaction.
The activity of this enzyme appears to be in the range of 0.027
umolar units/ml of erythrocytes. Recently, Meyskens and Williams
(1971) have reported that the K_m for adenosine for this enzyme is
1.9×10^{-6} M. If adenosine were to enter the erythrocyte in
relatively high concentrations (as during tissue breakdown), the
degradative adenosine deaminase reaction would be expected to predominate. On the other hand, if adenosine entered the erythrocyte
in relatively low concentrations, e.g., 1×10^{-6} M, as the result

of normal tissue adenine nucleotide turnover, the salvage or adenosine kinase reaction would predominate. Thus it seems likely that the interrelation between the Michaelis constants of these enzymes may play a critical role in determining the salvage or degradation of adenosine by human erythrocytes. Of course, it must also be borne in mind that allosteric regulation of these enzymes may play a significant role. A number of these conclusions have been arrived at independently by Meyskens and Williams (1971).

The conclusion drawn from pH studies with HGPRTase that the unionized form of the purine ring is the effective substrate is similar to that of earlier studies of UDPG dehydrogenase with the analog nucleotides 5-F-UDPG and 6-azaUDPG (Goldberg et al., 1963). These observations suggest the possibility that certain of the purine analogs may function more efficiently in an acidic intracellular environment. The availability of HPLC and recent extension of this technology to include the determination of nucleotides of thiolated purines (Nelson et al., 1973) should facilitate such studies.

REFERENCES

Agarwal, K. C. and Parks, Jr., R. E. (1972), Mol. Pharmacol. 8, 128

Agarwal, R. P. Sagar, S. M. and Parks, Jr., R. E. (1973), Fed. Proc. 32, 512

Agarwal, R. P., Scholar, E. M., Agarwal, K. C. and Parks, Jr., R. E. (1971), Biochem. Pharmacol. 20, 1341

Brown, P. R. (1970), J. Chromatog. 52, 257

Brown, P. R. (1973), "High Pressure Liquid Chromatography", Academic Press, N. Y.

Brown, P. R., Agarwal, R. P., Gell, J. and Parks, Jr., R. E. (1972), Comp. Biochem. Physiol. 43B, 891

Brown, P. R. and Parks, Jr., R. E. (1973), this volume

Dixon, M. (1952), Biochem. J. 55, 161

Goldberg, N. D., Dahl, J. L. and Parks, Jr., R. E. (1963), J. Biol. Chem. 238, 3109

Horvath, C., Preiss, B. and Lipsky, S. (1967), Anal. Chem. 39, 1422

Krenitsky, T. A., Papaioannou, R. and Elion, G. B. (1969), J. Biol. Chem. 244, 1263

Manohar, S. V., Lerner, M. H. and Rubenstein, D. (1968), Can. J. Biochem. 46, 455

Meyskens, F. L. and Williams, H. E. (1971), Biochim. Biophys. Acta 240, 170

Miller, R. L. and Bieber, A. L. (1968), Biochemistry 7, 1420

Miller, R. L. and Bieber, A. L. (1969), Biochemistry 8, 603

Moore, E. C. and LePage, G. A. (1958), Cancer Res. 18, 1075

Nelson, D. J., Bugge, C. J. L., Krasny, H. C. and Zimmerman, T. P. (1973), J. Chromatog. 77, 181

Parks, Jr., R. E. and Agarwal, R. P. (1972), In "The Enzymes", Vol. 7, 3rd ed. (Paul D. Boyer, ed.), Academic Press, N. Y.

Parks, Jr., R. E. and Brown, P. R. (1973), Biochemistry (in press)

Parks, Jr., R. E., Brown, P. R., Cheng, Y-C., Agarwal, K. C., Kong, C. M., Agarwal, R. P. and Parks, Christopher C. (1973), Comp. Biochem. Physiol. 45B, 355

Ross, A. F., Agarwal, K. C., Chu, S.-H. and Parks, Jr., R. E. (1973), Biochem. Pharmacol. 22, 141

Scholar, E. M., Brown, P. R., Parks, Jr., R. E. and Calabresi, P. (1973), Blood (in press)

Way, J. L. and Parks, Jr., R. E. (1958), J. Biol. Chem. 231, 467

This work was supported by grants: ACS-IC62 from the American Cancer Society, CA-07340 of the National Cancer Institute, USPHS and GMS 16538 of the National Institute of General Medical Science, USPHS.

A ROLE OF LIVER ADENOSINE IN THE RENEWAL OF THE ADENINE NUCLEOTIDES OF HUMAN AND RABBIT ERYTHROCYTES

Bertram A. Lowy and Marvin H. Lerner

Department of Biochemistry

Albert Einstein College of Medicine, New York, N.Y. 10461

For many years the work of our laboratory has been concerned with the synthesis and metabolism of the purine nucleotides in the mature human and rabbit erythrocyte and in the rabbit reticulocyte. Our early studies (1-5) and those of Mager's group (6,7) and others (8) have demonstrated the metabolic renewal of the purine nucleotides within the circulating erythrocyte during its lifespan. The observed turnover occurs in the absence of the capacity for the de novo biosynthetic pathway in the erythrocytes of both species, although the cells do contain purine nucleoside phosphorylase as well as adenine phosphoribosyltransferase and hypoxanthine-guanine phosphoribosyltransferase. However, the human erythrocyte lacks adenylosuccinate synthetase (9) and thus is unable to convert IMP to AMP, an enzymatic capacity possessed by the rabbit red cell. The erythrocytes of both species do possess adenylosuccinase and can, therefore, convert adenylosuccinate to AMP, and succinylaminoimidazolecarboxamide ribotide to aminoimidazolecarboxamide ribotide (AICAR). Formylation of AICAR and ring closure, which can occur in the erythrocyte, then leads to IMP formation. We had reported some years ago that the loss of a portion of the de novo pathway accompanies the maturation of the rabbit reticulocyte (4). More recent studies have indicated that many of the enzymes required for the early steps of the pathway are lacking in the erythrocytes of both species (10).

In order to explain the maintenance of purine nucleotide levels and the known turnover of these compounds during the erythrocyte lifespan, it is necessary to invoke the salvage pathway. For the human erythrocyte, only adenine and adenosine can serve as adenine nucleotide precursors,

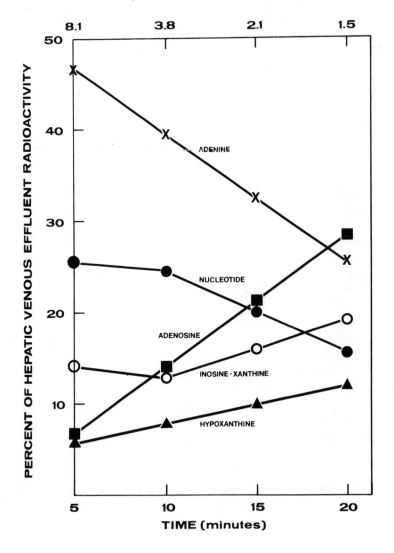

$[^3\text{H}]$ -ADENINE PERFUSION
Total Radioactivity In Acid–Soluble Extract (cpm X10^{-6})

the latter by the adenosine kinase reaction (11).

We have now utilized the technique of in situ perfusion for a study of the possible role of the liver in providing a purine or purine derivative to serve as a direct precursor of the purine nucleotides of the erythrocyte, as postulated by a number of investigators (12-16).

Rabbit livers, perfused free of blood, were then labeled by perfusion, in situ, via the portal vein, with a suitable purine nucleotide precursor: tritium labeled adenine or hypoxanthine. The livers were then perfused, with an isotonic solution containing a carrier nucleoside, usually unlabeled adenosine. Hepatic venous effluent was collected, during five minute periods, for twenty minutes. Cold dilute perchloric acid extracts were prepared and the purine components eluted by ion exchange chromatography on Dowex 50. The data obtained, following labeled adenine perfusion and subsequent perfusion with unlabeled adenosine are presented in Figure 1. The bulk of the radioactivity in the hepatic venous effluent, after 5 minutes, was associated with the adenine, a smaller percentage of radioactivity with the nucleotide fraction, and still smaller amounts with the inosine plus xanthine, adenosine and hypoxanthine fractions. At the end of 20 minutes, there was a dramatic decline in percent radioactivity associated with the adenine fraction, and a concomitant rise in the radioactivity of the adenosine fraction. The other fractions exhibited less pronounced changes.

In a similar experiment, a liver was labeled with $\left[^3H\right]$-hypoxanthine, followed by perfusion with carrier adenosine (Figure 2). The largest amount of radioactivity in the hepatic venous effluent after five minutes of un-labeled adenosine perfusion was associated with the inosine plus xanthine fraction, followed by hypoxanthine. As in the previous experiment, very little label appeared in the adenosine at this time point. At the end of twenty minutes, however, the radioactivity associated with adenosine had increased markedly, at the expense, although indirectly, of the labeled hypoxanthine. It is important to note that no detectable labeling of free adenine occurred.

The distribution and specific radioactivities of the purine compounds found in perchloric acid extracts prepared from the labeled livers, after the 20 minute carrier perfusion, were then determined. The data of Table 1 confirm the rapid uptake of labeled purine and conversion to nucleotide, presumably via the phosphoribosyltransferases of the cell. The specific activity data also support the initial labeling of the nucleotides and the values are consistent with the relatively large pool of nucleotides within the cell. In each experiment, significant labeling of adenosine, inosine

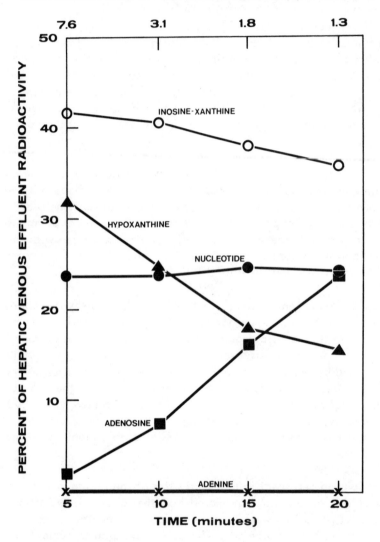

$[^3H]$ -HYPOXANTHINE PERFUSION
Total Radioactivity In Acid-Soluble Extract (cpm X10^{-6})

TABLE 1

Distribution of radioactivity and specific activity of acid-soluble purine compounds of perfused rabbit livers.

Rabbit livers were labeled with either [3H]-adenine or [3H]-hypoxanthine by perfusion in situ. The livers were subsequently perfused as indicated for a period of twenty minutes. Cold dilute perchloric acid extracts of the total livers prepared and analyzed for purine compounds by ion exchange chromatography on Dowex 50. Fractions were rechromatographed for specific activity on paper chromatography or electrophoresis.

Purine Compound	Labeled with [3H]-adenine		Labeled with [3H]-hypoxanthine			
	Perfused with adenosine (1.5mM)		Perfused with adenosine (1.5mM)		Perfused without carrier	
	Percent of total radio-activity	cpm/μmole	Percent of total radio-activity	cpm/μmole	Percent of total radio-activity	cpm/μmole
* Nucleotides	78	141,100	80	110,000	74	126,100
Adenosine	7	58,100	3	30,800	2	62,100
Inosine	6**	53,200	9**	35,500	12**	58,100
Hypoxanthine	3	44,000	8	40,400	12	80,400
Adenine	6	>500,000	<1	***	<1	***
Total radioactivity	cpm X10⁻⁸ 1.3		cpm X10⁻⁸ 1.0		cpm X10⁻⁸ 1.1	

* Specific activity calculation based on molar extinction coefficient of 14.7×10^3 for AMP at 257nm.

** Fraction contains 10% xanthine. Rechromatographed inosine was free of xanthine.

*** Negligible radioactivity and negligible free adenine.

TABLE II

Distribution of radioactivity in purine nucleotides of rabbit liver and human erythrocytes following perfusion of [^{3}H]-hypoxanthine labeled rabbit liver with washed human erythrocytes.

Rabbit liver was perfused with [^{3}H]-hypoxanthine, as previously described, followed by washout perfusion for ten minutes with an isotonic solution of salts containing glucose, to remove extracellular label. An oxygenated washed human erythrocyte suspension was then perfused through the liver for 1 hr. Erythrocytes were collected and washed in an isotonic solution of salts. Liver and erythrocyte extracts were chromatographed and assayed. Purine bases, prepared by hydrolysis of the nucleotide fractions, were separated by chromatography on Dowex 50.

Tissue	Purine nucleotide	Percentage of total nucleotide radioactivity
Rabbit liver	Adenine	49.5
	Guanine	9.0
	Hypoxanthine + Xanthine	41.5
Human erythrocyte	Adenine	82.5
	Guanine	8.0
	Hypoxanthine + Xanthine	9.5

and hypoxanthine occurred. In the adenine labeling experiment, the high radioactivity of the minute amount of free adenine isolated was probably an indication of precursor not yet converted to AMP (Figure 1).

These experiments strongly suggest that rabbit livers, perfused with either adenine or hypoxanthine, rapidly take up the purine, which is converted to nucleotide intracellularly. Specific activity data for the individual nucleotides, to be presented elsewhere, provide evidence for the known interconversions, in the liver, of AMP, IMP, XMP and GMP. The data presented here, indicate that conversion of AMP to adenosine occurs, probably as a result of the 5'-nucleotidase of the liver cell membrane (17-19). The adenosine, formed at the membrane, is then released from the cell, perhaps in a manner analogous to that described by Berne et al, in their studies of the myocardial cell (20,21).

In order to determine whether the released adenosine could serve as a precursor of human erythrocyte adenine nucleotides, a rabbit liver was labeled by perfusion with [3H] -hypoxanthine as described. Washout perfusion with an isotonic balanced salt solution removed residual extracellular label. The labeled liver was then perfused by recirculating a 400ml washed human erythrocyte suspension for one hour. Erythrocytes were then collected and washed, and the liver was excised. Extracts were prepared and the purine nucleotides were assayed for distribution of radioactivity (Table II). Within the liver, the radioactivity was approximately evenly distributed between the adenine nucleotides and the hypoxanthine plus xanthine nucleotides, indicating again that extensive conversion of hypoxanthine to IMP to AMP can occur in the liver cell. Within the human erythrocyte, over 80 percent of the label appeared in the adenine nucleotides. Since IMP is not converted to AMP in that cell, the labeled adenosine formed in the liver from the perfused hypoxanthine must have been taken up by the human erythrocyte and converted to AMP by the adenosine kinase. Since free adenine, the only other possible precursor of human erythrocyte adenine nucleotides, was not detected in hypoxanthine perfused liver or in hepatic venous effluent from hypoxanthine perfused liver, a possible role is unlikely.

In an extension of the erythrocyte perfusion experiment, a liver labeled by hypoxanthine perfusion, was removed from the rabbit and slices were prepared. The labeled liver slices were incubated for one hour in suspensions of washed human and rabbit erythrocytes, in vitro. The results, presented in Table III, indicate that as in the in situ experiment, over 80 percent of the nucleotide label in the human erythrocyte was present in the adenine nucleotides, undoubtedly derived from adenosine released from the liver cells. It is interesting to note that a similar distribution of label appeared in the

TABLE III

Transfer of radioactivity from $[^3H]$-hypoxanthine labeled rabbit liver to human and rabbit erythrocyte purine nucleotides, in vitro.

Rabbit liver was perfused with $[^3H]$-hypoxanthine as described. Slices prepared from the liver were washed with an isotonic solution of salts and then incubated in a washed human or rabbit erythrocyte suspension for 1 hour at 37°C, under 95% O_2–5% CO_2. The erythrocytes and tissue slices were separated, washed and analyzed.

Tissue	Percent purine radio-activity transferred from liver	Purine nucleotide	Radioactivity after 1 hr of incubation	Percentage of total erythrocyte nucleotides
			c pm	
Human erythrocyte	9.6	Adenine	32,900	87.1
		Guanine	2,060	5.5
		Hypoxanthine + Xanthine	2,780	7.4
Rabbit erythrocyte	10.1	Adenine	35,120	83.2
		Guanine	2,830	6.7
		Hypoxanthine + Xanthine	4,260	10.1

nucleotides of the rabbit erythrocyte although that cell does have the enzymic capacity to convert hypoxanthine to AMP, via IMP. It may well be that the rabbit erythrocyte also utilizes liver adenosine as an adenine nucleotide precursor, but hypoxanthine and/or inosine may also be utilized.

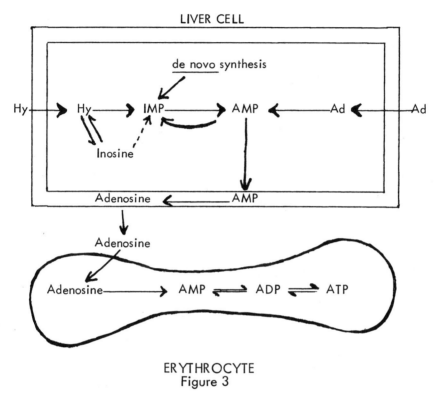

LIVER CELL

ERYTHROCYTE
Figure 3

The results of this investigation are summarized in Figure 3. Rabbit liver adenine nucleotides formed as a consequence of in situ perfusion with adenine or hypoxanthine in these experiments, were converted to adenosine, probably at the cell membrane by a 5'-nucleotidase. The adenosine released from the cell was rapidly taken up by the circulating erythrocyte. Within the erythrocyte, phosphorylation to AMP occurred, thereby suggesting a possible mechanism for the maintenance and renewal of the adenine nucleotides in the absence of biosynthesis via the de novo pathway, and in the human cell, in the absence of adenylosuccinate synthetase.

This investigation was supported by grants from the National Institutes of Health (AM-02655) and the American Heart Association (70-673).

Bibliography

1. Lowy, B.A., Ramot, B. and London, I.M. (1960) J. Biol. Chem. 235, 2920-2923.

2. Lowy, B.A. and Williams, M.K. (1960) J. Biol. Chem. 235, 2924-2927.

3. Lowy, B.A., Williams, M.K. and London, I.M. (1961) J. Biol. Chem. 236, 1439-1441.

4. Lowy, B.A., Cook, J.L. and London, I.M. (1961) J. Biol. Chem. 236, 1442-1445.

5. Lowy, B.A., Williams, M.K. and London, I.M. (1962) J. Biol. Chem. 237, 1622-1625.

6. Mager, J., Dvilansky, A., Razin, A., Wind, E. and Izak, G. (1966) Israel J. Med. Sci. 2, 297-301.

7. Mager, J., Hershko, A., Zeitlin-Beck, R. Shoshani, T. and Razin, A. (1967) Biochim. Biophys. Acta 149, 50-58.

8. Bishop, C. (1961) J. Biol. Chem. 236, 1778-1779.

9. Lowy, B.A. and Dorfman, B. (1970) J. Biol. Chem. 245, 3043-3046.

10. Lowy, B.A. and McDonagh, E. unpublished observations.

11. Lowy, B.A. and Williams, M.K. (1966) Blood 27, 623-628.

12. Smellie, R.M.S., Thomson, R.J. and Davidson, J.N. (1958) Biochim. Biophys. Acta 29, 59-74.

13. Lajtha, L.G. and Vane, J.R. (1958) Nature 182, 191-192.

14. Henderson, J.F. and LePage, G.A. (1959) J. Biol. Chem. 234, 3219-3223.

15. Pritchard, J.B., Cnavez-Peon, F. and Berlin, R.D. (1970) Am. J. Physiol. 219, 1263-1267.

16. Syllm-Rapaport, I., Jacobasch, G., Prehn, S. and Rapaport, S. (1969) Blood 33, 617-627.

17. Busch, E.W., VonBorcke, I.M. and Martinez, B. (1968) Biochim. Biophys. Acta 166, 547-556.

18. Song, C.S., Kappas, A. and Bodansky, O. (1969) Ann. N.Y. Acad. Sci. 166, 565-573.

19. Widnell, C.C. and Unkeless, J.C. (1968) Proc. Nat. Acad. Sci. U.S.A. 61, 1050-1057.

20. Katori, M. and Berne, R.M. (1966) Circulation Res. 19, 420-425.

21. Rubio, R., Berne, R.M. and Katori, M. (1969) Am. J. Physiol. 216, 56-62.

PURINE NUCLEOSIDE METABOLISM IN ESCHERICHIA COLI

Bjarne Jochimsen

Institute of Biological Chemistry B

University of Copenhagen, Denmark

This manuscript will mainly deal with purine nucleoside metabolism, and I would like to stress that the organism studied is the enteric bacterium Escherichia coli, with its natural habitat:the human rectum.

On figure 1 the interconversions of purine compounds are shown, as they have been established for the Enterobacteriaceae. The lower part of this figure is the one of interest for this communication.

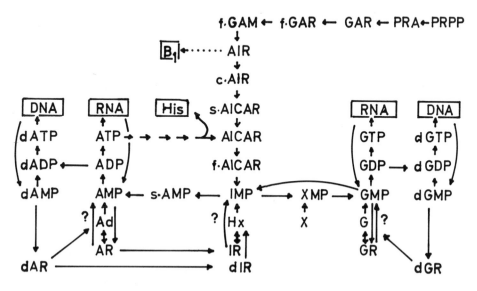

figure 1, Purine metabolism in Enterobacteriaceae.

Figure 2 shows the possible interconversions between purine bases, nucleosides and nucleoside monophosphates. From genetic and enzymatic studies it has been
shown that the conversion of the free bases(adenine,
hypoxanthine,xanthine and guanine) to the monophosphate
level is carried out by at least three different enzymes (the phosphoribosyltransferases)(1,6). This is in
contrast to human tissues, which have one specific adenine phosphoribosyltransferase and one transferase with
activity towards hypoxanthine,xanthine and guanine.

In contrast, it has been shown, that the phosphorolytic cleavage of the three nucleosides: adenosine to
adenine, inosine to hypoxanthine and guanosine to guanine, is carried out by one single enzyme. The existence of only one nucleoside phosphorylase has been demonstrated, both by enzymatic and genetic studies (8,5).

Thus the utilization of exogenous purine nucleosides involves three phosphoribosyltransferases and one
nucleoside phosphorylase, if a phosphorolytic cleavage
of the nucleosides is the first reaction to take place.
An alternative pathway for the utilization of nucleosides would be a direct phosphorylation to monophosphates.
The question is then: Do such nucleoside kinases exist
in E.coli,and if this is the case, how many enzymes are
involved?

Using a purine requiring mutant, defective in nucleoside catabolism (lacking nucleoside phosphorylase

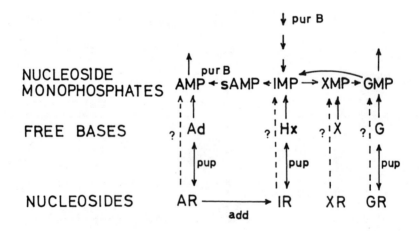

figure 2, Metabolism of exogenous purine bases and
nucleosides. Genesymbols: pur B, pup, add indicate genes
coding for adenylosuccinate lyase, purine nucleoside
phosphorylase, and adenosine deaminase respectively.

and adenosine deaminase; the reactions given by gene
symbols in figure 2) it has been possible to demonstrate the in vivo existence of kinase activities towards
adenosine, inosine and guanosine (4). The pathways operating in this mutant are given in figure 3.

The mutant strain (SØ 405) lacks the enzyme adenylosuccinate lyase, required for de novo purine biosynthesis as well as for the conversion of s-AMP to AMP.
This strain therefore has a specific requirement for an
adenine compound and a general purine requirement. As
the strain lacks nucleoside phosphorylase and adenosine
deaminase the pathways for nucleoside utilization via
free bases are ruled out.

In this mutant, adenosine, inosine or guanosine, in
combination with a free base, can fullfill the purine
requirement, indicating the presence of functional nucleoside kinases.

Figure 3 gives the growth kinetics for this mutant.

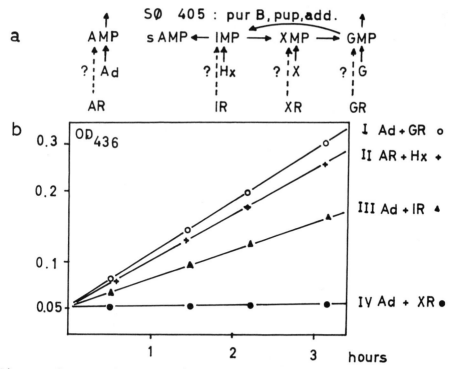

figure 3a, Pathways still operating in mutant SØ 405.
3b, Growth curves showing increase in optical density
(OD_{436}) versus time for the mutant grown on the indicated combinations of bases and nucleosides.

Separation of the three kinase activities, using several techniques, were tried in order to find out how many enzymes are involved?

Gelfiltration on a Sephadex G-100 column (figure 1) showed that none of the three activities were eluted together - this indicates the existence of three different enzymes.

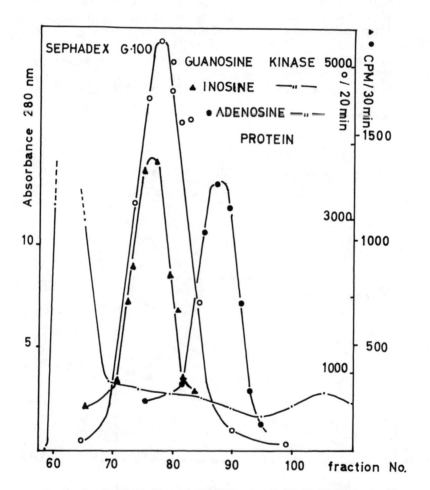

figure 4, Gelfiltration on a Sephadex G-100 column (2,5 x 87 cm), showing enzymatic activity for the three nucleosides versus fraction number (2,8 ml/fraction). Elution buffer: 20mM Tris-Cl pH 7.8, 20mM KCl, 5mM $MgCl_2$, 1mM mercaptoethanol. Flow rate 16 ml/hour. Assays according to reference 10.

It was then of interest to see, whether the three
kinases were genetically separable (i.e., will a loss
in one kinase activity affect the two others)?

For this purpose, we selected mutant derivatives
in SØ 405, unable to grow on nucleosides as purine
source. Two types of mutants were expected to result in
the desired phenotype. i) mutants lacking the nucleoside kinase, and ii) mutants unable to take up nucleosides.

Of the first class, only one mutant, was obtained,
lacking guanosine kinase (<1%), and this mutant is also
lowered in inosine kinase (30% of parent), while adenosine kinase is normal. This result does not clarify,
the question, whether the kinases are genetically separable.

Of the second class, a mutant, unable to take up
adenosine, and a mutant, unable to take up guanosine
and inosine were obtained. When tested by assay on cell
extracts these mutants showed normal kinase activities
for all three nucleosides.

Uptake (incorporation) curves for these mutants are
given in figure 5. Very little is at present known about
uptake of purine nucleosides in E.coli, but these data
indicate that E.coli, in addition to the three kinases,
contains at least two other proteins necessary for transport (uptake) of purine nucleosides.

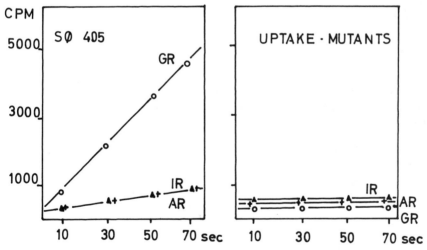

figure 5, Uptake (incorporation) of nucleosides in parent (SØ 405) and uptake mutants. Cells were starved,
and at time zero, ^{14}C-(U)-nucleosides (spec.act. 100 uCi
/u mole) were added to a final concentration of 0.1 uM.
Samples (200 ul) were withdrawn at the times indicated,
and filtered rapidly (and washed) on millipore filters.

In conclusion, it can be said that at least 10 enzymes (proteins) are involved in purine salvage. Figure
6 gives a short summary of our present knowledge about
purine salvage in enteric bacteria.

AUXILIARY PATHWAYS   (PURINES)

Enzymatic activity	number of enzymes	REGULATION
PHOSPHORIBOSYL TRANSFERASES	3	Induction: BASES (3)
NUCLEOSIDE PHOSPHORYLASE	1	—— „ —: NUCLEOSIDES (2,7)
NUCLEOSIDE KINASES	3	?
NUCLEOSIDE UPTAKE	min 2	?
ADENOSINE DEAMINASE	1	Induction: BASES (9)

figure 6, Summary of enzymes (and their regulation)
involved in purine salvage.

REFERENCES

1. Gots,J.S. J.Bacteriol.112(1972)910-916
2. Hammer-Jespersen,K. Eur.J.Biochem.19(1971)533-538
3. Hochstadt-Ozer,J. J.Biol.Chem.246(1971)5294-5303
4. Hoffmeyer,J. J.Bacteriol.106(1971)14-24
5. Karlström,O. J.Bacteriol.95(1968)1069-1077
6. Krenitsky,T.A. J.Biol.Chem.245(1970)2605-2611
7. Munch-Petersen,A. Eur.J.Biochem.6(1968)432-442
8. Nygaard,P. Personal communication
9. Remy,C.N. J.Bacteriol.96(1968)76-85
10. Assays, purine nucleoside kinase assays were made
by mixing equal volumes of fraction and reaction mixture:
(0,14 M Tris-Cl pH 7.8 (6.5),0,07 M $MgCl_2$, 0,07 M KCl,
2mM ATP and 1mM ^{14}C-(U)- nucleoside (50.000 cpm/ml).
Reaction mixture plus enzyme was incubated at $37^\circ C$ for
20 (30) minutes and 15 μl samples were taken and spotted on poly-ethylene-imine coated plastic sheets. Plates
were developed with water, and start spots containing
nucleotides were cut out, and counted.

SYNTHESIS OF PURINE NUCLEOTIDES IN HUMAN AND LEUKEMIC CELLS. INTERACTION OF 6-MERCAPTOPURINE AND ALLOPURINOL.[1]

W.Wilmanns

University Clinic, Department of Internal

Medicine II, Tuebingen - Germany

In proliferating cells, purine nucleotides - necessary for DNA- and RNA-synthesis - can be formed by incorporation of small precursors (de novo pathway) and by utilisation of free purin bases (salvage pathway). The most important reactions of both metabolic pathways are demonstrated in figure 1. The first reaction of the de-novo pathway - the formation of phosphoribosylamine from glutamine and phosphoribosylpyrophosphate (PRPP) by the enzyme PRPPamidotransferase - is the target of a negative feed back control mechanism by the endproducts IMP,GMP and AMP (1).

In a similar way the purine antimetabolite 6-mercaptopurine (6-MP), after conversion to the monophosphoribotide, acts as an inhibitor of this enzyme (9.10). There have been described other sites of action of 6-MP (4-7) which are marked by arrows in figure 1. We have previously demonstrated the inhibition of the enzymic formate activation (tetrahydrofolate formylase) in leukemic cells by 6-MP (12). Inspite of a rather high inhibitory concentration of 6-MP between 10^{-3} and 10^{-4} M, this inhibition has some practical clinical implications for the treatment of acute leukemia, as it is detected in sensitive leukemic cells only. In accordance with reports of several authors (8,9,10) we have postulated, that 6-MP has to be converted into 6-thioinosinic acid for exerting its inhibitory effect on the de novo synthesis of purinenucleotides. On the other hand DAVIDSON and WINTER have shown that both 6-MP sensitive and resistant cells con-

1) With the aid of Deutsche Forschungsgemeinschaft
   (Forschergruppe Leukaemie- und Tumortherapie)

─── inhibition by 6-MP

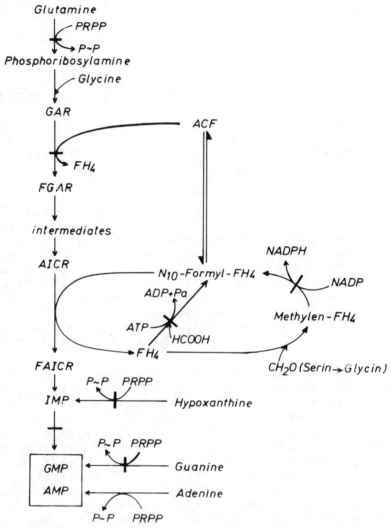

Figure 1
Purine nucleotide synthesis and its inhibition by 6-mer-
captopurine (6-MP).PRPP:phosphoribosylpyrophosphate, GAR
glycinamidribotide,FGAR formylglycinamidribotide, AICR:
5-amino-imidozolcarboxamide-ribotide,FAICR:5-formamino-
4-imidazolcarboxamide-ribotide,IMP:inosine-5'-monophos-
phate,AMP:adenosine-5'-monophosphate,GMP:guanosine-5'-
monophosphate,FH_4:tetrahydrofolate,HCOOH:formate

tain a purine phosphoribosyltransferase, which is specific for hypoxanthine, guanine and 6-MP (2). Moreover, UNGER
and SILBER did not find any marked difference between the
inhibitory effect of 6-MP and thioinosinic acid on the enzymic activation of formate (11).

Allopurinol is known to potentiate the cytocidal effect
of 6-MP on cell proliferation. This is explained by inhibition of xanthine oxidase. By this enzyme not only hypoxanthine is catabolized to uric acid but also 6-MP to
thiouric acid.

The carbon atoms 2 and 8 in the purine ring of inosinic
acid are derived from "C1" units. The latter are transferred as activated formate to GAR and AICR as specific
formate acceptors. Therefore we have studied the tetrahydrofolate dependent activation of formate in relation
to the netto de novo synthesis of purine nucleotides in
cell-free extracts of normal and leukemic leukocytes. In
addition, the conversion of exogenous purines to corresponding monophosphoribonucleotides by the specific purine-
phosphoribosyltransferases was determined. The aim of these
investigations was to study the effect of 6-MP on the formate activating system, which is important for the de novo
synthesis of purine nucleotides, on the utilization of
preformed purine bases and, in addition, the interaction
of allopurinol with 6-MP.

METHODS

The details of cell preparations and the assay of the
tetrahydrofolate formylase are described elsewhere (12).

For determination of the tetrahydrofolate formylase activity sodium formate and tetrahydrofolate were incubated
with ATP, ATP-regenerating system and Mg^{++} ions in TRIS-
puffer pH 7.4 60 minutes at $37^{\circ}C$. By stopping the reaction
with perchloric acid all N10-formyltetrahydrofolic acid
is converted to anhydrocitrovorum factor (ACF) which was
determined by measuring the absorption at 355 m/u in a
Zeiss Spectrophotometer (ε =22.000).

For comparising the activity of the tetrahydrofolate formylase with the netto de novo synthesis of purine nucleotides incubations were performed with 14C-formate and the
components necessary for the enzymic formate activation
only and in a parallel experiment with additional $KHCO_3$,
glutamine,glycine and ribose-5-phosphate (12). Determinations of the tetrahydrofolate formylase activity and of

the amount of purine nucleotides synthesized de novo
were performed by measuring the fixation of 14 C-formate
into acid insoluble ACF and purine nucleotides. The latter
were identified after hydrolysis with HCl by paper chromatography (fig.2).

For the assay of purine phosphoribosyltransferase the
incubation mixture contained in a volume of o.5 ml:

        20 /umol TRIS pH 7.6

        0.25 /umol 14C-purine (adenine, hypoxanthine,
                            guanine or 6-MP) (spec.
                            activity ca.
                            500 000 cpm/ /umol)

        1 /umol phosphoribosylpyrophosphate (stabilized
                                     with EDTA)

        cell-free extracts from 2 x 10^{+6} leukocytes or
                                       leukemic cells.

After 1 hour incubation at 37°C the reaction was termi-
nated by heat inactivation. The free purine base and the
corresponding purinemonophosphonucleotide were seperated
by paper chromatography, using n-butanol/glacial acetic
acid/H_2O = 12/3/5 as solvents (fig.7). The radioactivity
was measured in a Packard Liquid Scintillation Spectrometer.

RESULTS

As is illustrated in figure 2 incorporation of 14C-for-
mate into purine nucleotides can be detected only, if
all components that are necessary for the de novo pathway are added to the incubation mixture together with
the substrates and cofactors of the formate activating
system. In this experiment the purines identified by
paper chromatography were hypoxanthine and adenine.

The tetrahydrofolate formylase was found in normal and
leukemic leukocytes, the highest activities in immature
blast cells of acute Leukemia (table 1). By determination
of the 14C-formate incorporation a measurable netto de
novo synthesis of purine nucleotides was only detected
in immature leukemic cells.

If different patients with acute leukemia were studied,
the amount of 14C-formate incorporation was correlated
to the activity of the formate activating enzyme as illustrated in figure 3. In addition, it was found that the
rate of 14C-formate incorporation into purine nucleotides depends on the concentration of tetrahydrofolate
(FH_4) in the incubation mixture (fig. 4).

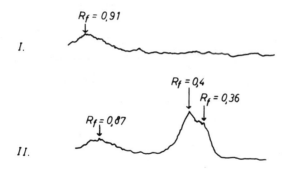

Figure 2

14C-formate incorporation into purines in cell-free extracts of peripheral blast cells of a patient with acute
leukemia
Components: I Components of the formate activating enzyme
            II In addition: $KHCO_3$, glutamine, glycine,ri-
               bose-5-phosphate
Analysis by paper chromatography (Schleicher a.Schüll
2043 Mgl;Isopropanol/H_2O/conc. HCl = 17/4.1/3.9)
R_f: Hypoxanthine o.36
     Adenine   o.4

Table 1

Activities of the formate activating enzyme and incorporation rates of 14C-formate into purine nucleotides in
cell-free extracts of normal leukocytes and bone marrow
cells and of leukemic cells.

Spez.Act.(μmol/min/10^{10}cells)

cells investigated	Formate activation	14C-Formate incorporation
normal peripheral leukocytes	0.38	o
	0.23	o
normal bone marrow cells	0.93	o
	0.55	o
peripheral blast cells of patients with acute leukemia	1.14	0.19
	1.67	0.61
	1.72	0.30
	1.84	0.57
	3.00	0.67
	2.52	0.90
	1.52	0.83
	0.10	0.09

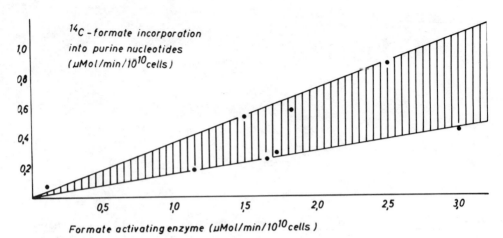

Figure 3
Correlation between the activities of the formate activating enzyme and the 14C-formate incorporation in leukemic cells of 8 patients with acute leukemia
r = 0.7067, P<0.05

Therefore it can be assumed, that the rate of purine nucleotide de novo synthesis is controlled not only by the
first reaction of this pathway, the PRPPamidotransferase,
but also by the amount of N_{10}-formyl-FH_4 available for
incorporation of activated formyl groups into the C_2 and
C_8 positions of the purine ring (fig.5).

In figure 6 the influence of 6-MP and allopurinol on the
tetrahydrofolate formylase in leukemic blast cells of 5
patients with acute leukemia is illustrated. 6-MP in a
rather high concentration of 1.5×10^{-3} M has only a small
inhibitory effect on this enzyme.By addition of allopurinol the inhibition of the enzyme by 6-MP was markedly
increased, whereas allopurinol alone had no effect. This
is explained by the fact, that allopurinol, by inhibition
of the xanthine oxidase, reduces the inactivation of 6-MP
to 6-thiouric acid.

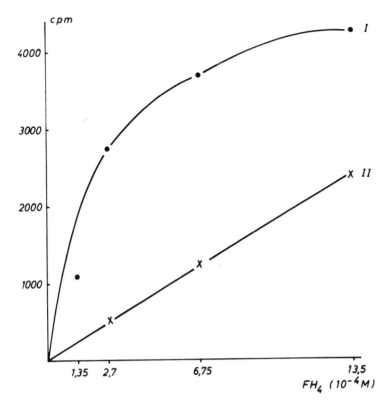

Figure 4
Incorporation of 14C-formate into Anhydrocitrovorum Factor (I) and into purine nucleotides (II) in relation to
the concentration of FH_4
Cell-free extract of blast cells of a patient with acute
myeloblastic leukemia

In figure 7 an autoradiogiagramm is shown, which demonstrates the chromatographic separation of 14C-hypoxanthine and
14C-IMP, formed by the hypoxanthine-guanine phosphoribosyltransferase, and the inhibition of this enzyme by 6-MP
(lower part). According to table 2 the 6-MP concentration,
necessary for inhibition of this reaction, is about 10times
lower as in the tetrahydrofolate formylase reaction. As
the enzyme hypoxanthine-guaninephosphoribosyltransferase
is also specific for 6-MP, this compound is transformed to
thio-inosinic acid, which is regarded by several authors
as the metabolically active inhibitor of the purine de
novo synthesis. The conversion of adenine to AMP is not
effected by 6-MP.

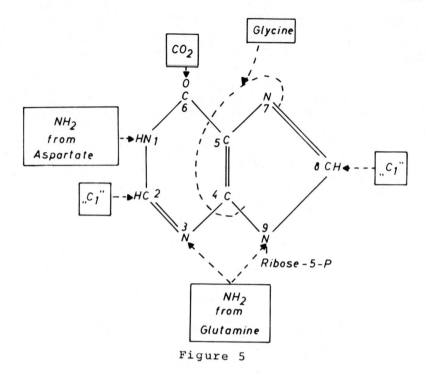

Figure 5

Precursors of IMP

Table 2

Effect of 6-MP on purine-phosphoribosyl-transferase
in cell-free extract of leukemic blast cells.

14C-Purine	Addition of 6-MP (1.5x10^{-4}M)	Enzyme Activity (μmol/min/10^{10} cells	14C-Nucleo-tide
Adenine	−	4.5	AMP
	+	5.1	
Hypoxanthine	−	4.3	IMP
	+	1.9	
Guanine	−	6.2	GMP
	+	2.5	
6−MP		4.1	Thio-IMP

Incubation: 14C-Purine   5 x 10^{-4}M
             PRPP   2 x 10^{-3}M   60' 37°C
             MgCl$_2$ 10 x 10^{-3}M   pH 7.6

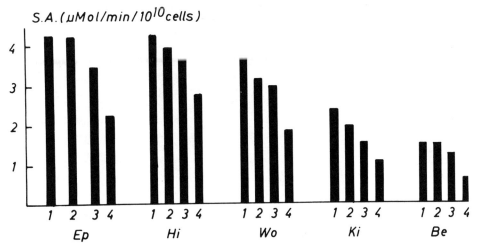

S.A.(µMol/min/10^{10}cells)

Figure 6
Activity of the formate activating enzyme in leukemic
blast cells of 5 patients with acute leukemia under the
action of 6-MP and allopurinol

1.Normal incubation
2.Addition of allopurinol (1.5 x 10^{-3}M)
3.Addition of 6-MP (1.5 x 10^{-3}M)
4.Addition of allopurinol and 6-MP

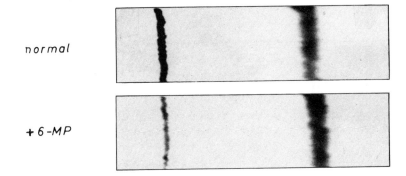

normal

+6-MP

Figure 7
Hypoxanthine-phosphoribosyl-transferase in leukemic cells

Chromatography:n-Butanol/glacial acetic acid/H$_2$O=12/3/5
  R$_f$ 14C-Hypoxanthine   0.58
        14C-IMP   0.08
  14C Incorporation into IMP: without 6-MP   44%
                             with   6-MP   22%

Wo. o, 27 years
AML

According to the experiments described in table 3 the
addition of allopurinol does effect neither the inhibition
of the hypoxanthine-guanine phosphoribosyltransferase by
6-MP nor the conversion of 6-MP to thioinosinic acid by
the same enzyme. It is unprobable that this is due to a
lack of xanthine oxidase in the cells investigated, because, in the same cell-free system the inhibitory effect
of 6-MP on the formate activation was markedly increased
by allopurinol (fig.6).

Table 3

Action of 6-MP and allopurinol on the
hypoxanthine-guanine-phosphoribosyltransferase

| 14C Substrate | Addition of | | Spez.Activity |
	6-MP 1.5×10^{-4} M	allopurinol 7.5×10^{-4} M	(μmol/min/10^{10} cells)
Guanine	-	-	10.3
	+	-	5.0
	+	+	4.6
Hypoxanthine	-	-	27
	+	-	11
	+	+	11
6 -mercaptopurine		-	13
		+	12

On the basis of experiments described it is concluded,
that a conversion of 6-MP to the monophosphoribonucleotide
is not necessary for the inhibition of formate activation.
Only the inhibitory effect of 6-MP on this reaction, which
is important for purine nucleotide de novo synthesis, is
potentiated by allopurinol. The competition of both substrates with the enzyme xanthine oxidase is assumed to be
the biochemical basis for this interaction.Thus, the inactivation of 6-MP by oxidation to thiouric acid can be
prevented by allopurinol. The clinical observation of a
higher cytotoxic effect of 6-MP-treatment, if patients re-
ceive allopurinol at the same time, is explained by these
results.

ACKNOWLEDGEMENTS

I wish to thank Miss Sabine Gebauer and Miss Erdmute
Schiller for their assistance with this study.

REFERENCES

1. CASKEY, C.T.,D.M.ASHTON, and J.B. WYNGAARDEN.(1964)
   The enzymology of feedback inhibition of glutamine
   phosphoribosylpyrophosphate amidotransferase by purine
   ribonucleotides. J.Biol.Chem. 239:2570

2. DAVIDSON,J.D.,and T.S. WINTER:Purine nucleotide
   Pyrophosphorylase in 6-Mercaptopurine-sensitive and
   resistant human leucemias. Cancer Res. 24 261 (1964)

3. ELION, G.B.,S.GALLAHAN, R.W.RUNDLES and G.H.HITCHINGS:
   Relationship between metabolic fates and antitumor
   activities of thiopurines. Canc. Res. 23 S.1207-1217
   (1963

4. GOTS, J.S., and J.GOLDSTEIN:Specific action of adenine
   as a feed back inhibitor of purine biosynthesis.
   Science 130, 622 (1959)

5. GOTS,J.S., and E.G.GOLLUP:Purine analogs as feedback
   inhibitors.Proc.Soc. exp.Biol. (N.Y.)101 641,(1959)

6. HENDERSON, J.F.:Feedback inhibition of purine biosynthesis in ascites tumor cells by purine analogues.
   Biochem. Pharmacol. 12, 551 (1963)

7. LE PAGE,G.A. and M.JONES:Purinethiols as feedback inhibitors of purine synthesis in ascites tumor cells.
   Cancer Res. 21, 642 (1961)

8. LUKENS,L.N., and K.A.HERRLINGTON:Enzymatic formation
   of 6-Mercaptopurine ribotide.Biochim.biophys. Acta
   (Amst.)24 432 (1957)

9. MC COLLISTER, R.J.,W.R. GILBERT jr.,D.M.ASHTON,and
   J.B.WYNGAARDEN:Pseudofeedback inhibition of purine
   synthesis by 6-Mercaptopurine ribonucleotide and
   other purine analogues. J.biol.Chem.239,1560 (1964)

10.MC COLLISTER,R.J.and J.B.WYNGAARDEN:Pseudofeedback of
   purine synthesis by 6-Mercaptopurine and other purine
   analogues.J.cl.Invest. 41, 1383 (1962)

11. UNGER,K.W., and R.SILBER:Studies on the formate acti-
    vating enzyme:Kinetics of 6-Mercaptopurine inhibition
    and stabilisation of the enzyme.Biochim.biophys.Acta
    (Amst.) 89, 167 (1964).

12. WILMANNS,W.:Formiat-Aktivierung und de novo-Synthese
    von Purin-Nucleotiden in Leukämiezellen und ihre
    therapeutische Beeinflussung durch 6-Mercaptopurin.
    Zschr. ges.exp. Med. 147, 154 (1968)

ENZYMES OF PURINE METABOLISM IN PLATELETS: PHOSPHORIBOSYLPYROPHOSPHATE SYNTHETASE AND PURINE PHOSPHORIBOSYLTRANSFERASES

Z. Jerushalmy, O. Sperling, J. Pinkhas
M. Krynska and A. de Vries

Rogoff-Wellcome Medical Research Institute and
the Metabolic Unit of Department of Medicine D,
Tel-Aviv University Medical School, Beilinson
Hospital, Petah Tikva, Israel

Human platelets do not synthesize nucleotides de novo (1) but have been demonstrated to possess salvage pathways utilizing adenine and adenosine (2,3). On the other hand, no evidence has been reported as yet for the presence of hypoxanthine-guanine phosphoribosyltransferase (HGPRT) in human platelets.

Nucleotide formation from free purine bases is dependent on the availability of phosphoribosylpyrophosphate (PRPP), the common substrate for both HGPRT and adenine phosphoribosyltransferase (APRT) (4). Whereas human platelets do contain PRPP, the activity of PRPP synthetase in these cells has not been reported thus far (2).

In this communication we report on the presence of both PRPP synthetase and HGPRT in human platelet lysates. The activity of these enzymes was also studied in rabbit platelets. Platelets were separated and washed according to Ardlie et al. (5) and suspended in tris-buffered saline pH 7.35 (3). The incorporation of purine bases into the nucleotides of intact platelets was investigated by incubating cell suspensions in tris-buffered saline containing labelled purine bases. The reactions were arrested by the addition of perchloric acid and the total nucleotides in the protein-free supernatant were separated from the purine nucleosides and bases by thin

layer chromatography (6). For the assay of the purine
phosphoribosyltransferases, cell suspensions were thrice
frozen and thawed in presence of 1 mM PRPP (7). The
lysates were centrifuged and the dialyzed supernatants
used for assay. For the assay of PRPP synthetase,
platelet suspensions in phosphate buffer were sonicated.
The lysates were centrifuged and the supernatant dialyzed
against phosphate buffer. PRPP synthetase was assayed
by coupling PRPP synthesis from ribose-5-phosphate and
ATP to formation of IMP by the addition of labelled
hypoxanthine and partially purified erythrocyte HGPRT.

It was found that human as well as rabbit platelet
lysates synthesize PRPP from ribose-5-phosphate and ATP.
The average specific activities of PRPP synthetase in
pooled human and pooled rabbit platelet lysates were 12
and 22 nmoles/mg protein/hr, respectively.

Both human and rabbit platelets were found to
possess, in addition to APRT (2) also HGPRT activity
(Table 1). The specific HGPRT and APRT activities
and the ratio between them are similar to those found in
cultured human skin fibroblasts (8), but differ from
those in human erythrocyte lysates in which the specific
activities of both enzymes are lower, especially that of
APRT (9). Human and rabbit platelets were also found to

TABLE 1. HGPRT AND APRT ACTIVITIES IN PLATELET LYSATES

Platelet source	Experiment	Nucleotides formed, nmoles/mg protein/hr		
		Hypoxanthine*	Guanine*	Adenine*
Human	1	162	222	214
	2	111	238	142
Rabbit	1	181	180	73
	2	226	221	102

* substrates

TABLE 2. INCORPORATION OF ^{14}C-LABELLED PURINE BASES INTO THE TOTAL NUCLEOTIDES OF INTACT PLATELETS

Platelet source	Experiment	Incorporation into nucleotides nmoles/mg protein/hr		
		Hypoxanthine	Guanine	Adenine
Human	1	0.20	0.40	26.0
	2	0.50	0.45	21.0
	3	0.40	1.70	15.0
	4	0.50	0.55	29.0
Rabbit	1	0.45	0.45	21.0
	2	0.25	0.50	26.0
	3	0.25	–	14.0

incorporate adenine into the total cell nucleotides. A much smaller incorporation of hypoxanthine and guanine was observed (Table 2). The present data on the incorporation of free purine bases into intact platelets are compatible with those of Holmsen and Rozenberg (2), only adenine being significantly accumulated in the nucleotide fraction. The greater accumulation of labelled adenine in the platelet nucleotides may be due to rapid phosphorylation of AMP to ATP (10), sparing the former from the hydrolytic action of the platelet nucleotidase (11). IMP, which in the human is not phosphorylated to IDP and ITP, would be prone to the nucleotidase activity, hydrolysing the IMP to hypoxanthine (11). This explanation might not hold for the presently observed failure of labelled guanine to accumulate in the nucleotide fraction of intact human and rabbit platelets, since at least for the rabbit platelet phosphorylation of GMP to GTP has been demonstrated (12).

References

1. Holmsen H. Scand. J. Clin. Lab. Invest.
    17:230-234, 1965.

2. Holmsen H. and Rozenberg M.C. Biochim. Biophys.
    Acta (Amst.) 157:266-279, 1968.

3. Holmsen H. and Rozenberg M.C. Biochim. Biophys. Acta
    155:326-341, 1968.

4. Wyngaarden J.B. and Kelley W.N. Gout: In Stanbury,
    J.B.; Wyngaarden J.B. and Fredrickson D.S.:
    The metabolic basis of inherited disease. (McGraw-
    Hill Book Co. New York, 1972) pp 905-917.

5. Ardlie, N.G., Packham M.A. and Mustard J.F.
    Brit. J. Haematol. 19:7-17, 1970.

6. Sperling O., Frank M., Ophir R., Liberman U.A.,
    Adam A. and de Vries A. Europ. J. Clin. Biol. Res.
    15:942-947, 1970.

7. Zoref E., Sperling O. and de Vries A. In press,
    Abst. of Internat. Symp. on Purine Metabolism in
    Man, Tel-Aviv, June 1973.

8. Zoref E., Sperling O. and de Vries A. In press.
    Abst. of Internat. Symp. on Purine Metabolism in
    man. Tel-Aviv, June 1973.

9. Kelley W.N., Green M.L., Rosenbloom F.M., Henderson
    J.F. Ann. Inter. Med. 70:155-206, 1969.

10. Holmsen H. and Day H.J. Ser. Haemat. 4:28-58, 1971.

11. Ireland D.M. Biochem. J. 105:857-867, 1967.

12. Da Prada M. and Pletcher A. Life Sci. 9:1271-1281,
    1970.

MUTATIONS AFFECTING
PURINE METABOLISM

Properties of HGPRT and APRT in HGPRT Deficient Blood Cells

HYPOXANTHINE-GUANINE PHOSPHORIBOSYLTRANSFERASE DEFICIENCY: ALTERED

KINETIC PROPERTIES OF A SPECIFIC MUTANT FORM OF THE ENZYME

J.A. McDonald and W.N. Kelley

Department of Medicine, Duke University Medical Center

Durham, North Carolina   27710

The Lesch-Nyhan syndrome is a rare, X-linked genetic disease due to a functional absence of the enzyme hypoxanthine-guanine phosphoribosyltransferase (HGPRT) (Seegmiller, Rosenbloom and Kelley, 1967). This enzyme catalyzes the transfer of the 5-phosphoribosyl moiety of 5-phosphoribosyl-1-pyrophosphate (PP-ribose-P) to the purine bases guanine and hypoxanthine to form the nucleotides inosinic acid and guanylic acid.

We have studied a patient with the classical Lesch-Nyhan syndrome (Fig. 1 and 2) who does not have a complete absence of

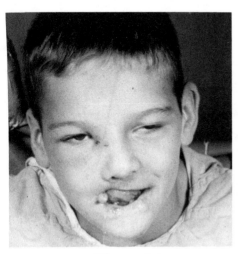

Figure 1

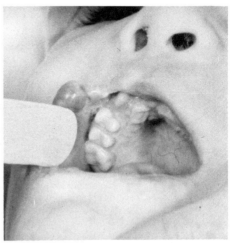

Figure 2

HGPRT but rather possesses a mutant form of HGPRT with normal activity but strikingly altered kinetic properties (McDonald and Kelley, 1971). This particular mutant form of HGPRT differs in several distinctive ways from the normal human enzyme. First, the apparent Km of the mutant HGPRT for the purine substrates hypoxanthine and guanine is 10-fold higher than that of the normal enzyme (Table 1). The apparent Km of the normal HGPRT for guanine is 5.0 X 10^{-6}M; that of the mutant HGPRT is 4.8 X 10^{-5}. The apparent Km of the normal HGPRT for hypoxanthine is 1.7 X 10^{-5}, while that of the mutant enzyme is 1.8 X 10^{-4}. Secondly, the mutant HGPRT requires greater than 10-fold higher concentrations of PP-ribose-P for maximal activity (Table 2). The apparent Km of the normal HGPRT for PP-ribose-P is 2.5 X 10^{-4}M assayed at saturating guanine. The mutant HGPRT is half-saturated by PP-ribose-P at 3.2 X 10^{-3}M at saturating guanine concentrations. The apparent Km of the mutant enzyme for PP-ribose-P using hypoxanthine as the constant purine substrate is 2.8 X 10^{-3}M, which represents an

TABLE 1

COMPARISON OF THE KINETIC CONSTANTS OF NORMAL AND MUTANT HYPOXAN-
THINE-GUANINE PHOSPHORIBOSYLTRANSFERASE FOR HYPOXANTHINE AND
GUANINE

Enzyme Source		Apparent	Apparent
Cell	Type	Km (guanine) (M)	Km (hypoxanthine) (M)
Erythrocyte	Normal	5.0X10^{-6}	1.7X10^{-5}
	Mutant	4.8X10^{-5}	1.8X10^{-4}

(From McDonald and Kelley, 1971)

TABLE 2

COMPARISON OF THE KINETIC CONSTANTS OF NORMAL AND MUTANT
HYPOXANTHINE-GUANINE PHOSPHORIBOSYLTRANSFERASE FOR PP-RIBOSE-P

Enzyme Source		Apparent
Cell	Type	Km (Mg PP-ribose-P) (M)
Erythrocyte	Normal	2.5X10^{-4}
	Mutant	3.2X10^{-3}*
Fibroblast	Normal	1.0X10^{-4}
	Mutant	2.0X10^{-3}*

* Values depict $S_{0.5}$ rather than Km.
   (From McDonald and Kelley, 1971)

increase in Km similar to that observed with guanine as the constant
purine substrate. Skin fibroblast preparations from the patient
also exhibited a similar increase in Km for PP-ribose-P from
1×10^{-4} to greater than 2×10^{-3}M.

The substrate kinetics of this mutant HGPRT also exhibit a
striking qualitative difference from the normal enzyme. As illustrated in Fig. 3, this mutant HGPRT has sigmoid kinetics with
PP-ribose-P as the variable substrate, whereas the normal HGPRT
exhibits hyperbolic, Michaelis-Menten kinetics with this substrate.
The inset is a magnified view of the values observed at the lower
range of PP-ribose-P concentration. This mutant HGPRT retained
Michaelis-Menten kinetics with either hypoxanthine or guanine as
the variable substrate. Using the Michaelis-Menten formula to
extrapolate to a saturating concentration of hypoxanthine, this
mutant HGPRT in hemolysates has virtually 100% of normal specific
activity when assayed at high PP-ribose-P levels (Table 3). Hill
plots derived from the rate versus PP-ribose-P curves were linear
for both normal and mutant HGPRT (Fig. 4). The slope of the Hill
plot for the mutant HGPRT was 2.3 consistent with the sigmoid nature
of the curve.

To further evaluate the differences between the normal enzyme

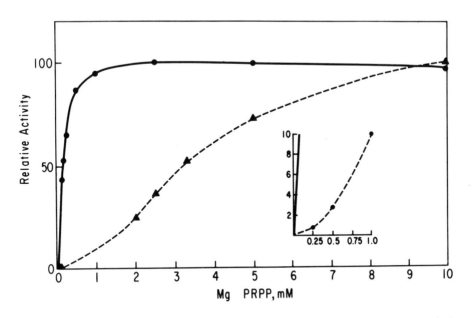

Fig. 3. Comparison of kinetic properties of normal and mutant hypoxanthine-guanine phosphoribosyltransferase. Normal enzyme, solid
line; mutant enzyme, dotted line (From McDonald and Kelley, 1971).

TABLE 3

COMPARISON OF V_{MAX} OF NORMAL AND MUTANT
HYPOXANTHINE-GUANINE PHOSPHORIBOSYLTRANSFERASE

Enzyme Source		Phosphoribosyltransferase Activity (nmole mg^{-1} hr^{-1})	
Cell	Type	Guanine	Hypoxanthine
Erythrocyte	Normal	98±14*	97±19*
	Mutant	12.1	94.4
Fibroblast	Normal	141±17**	
	Mutant	10.4	

*   Mean ± S.D. in 119 subjects
** Mean ± S.D. in 13 normal cell strains
  (From McDonald and Kelley, 1971)

and this mutant HGPRT with respect to PP-ribose-P binding, we
studied the effect of PP-ribose-P upon the rate of heat inactivation of the two different enzymes (Fig. 5).   The normal and mutant
HGPRT were heated at 80° for the period indicated in the presence
of varying amounts of PP-ribose-P.   Magnesium was kept at a
concentration equimolar to   that of   PP-ribose-P.   PP-ribose-P
at a   concentration   as low   as 1 X 10^{-5} was   found to afford
protection   of   the normal   HGPRT   against   heat denaturation.

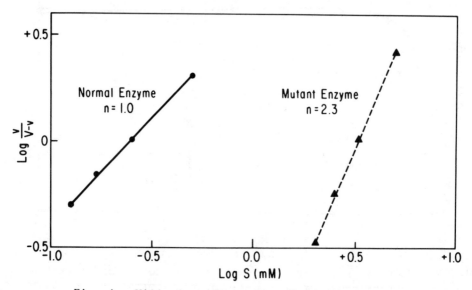

Fig. 4.   Hill plot of normal and mutant HGPRT.

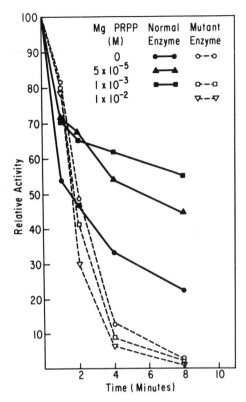

Fig. 5.   Influence of Mg PRPP on inactivation of normal and mutant PRT enzymes at 80°C (From McDonald and Kelley, 1971).

A higher concentration of PP-ribose-P, 1×10^{-3}M, provided even greater protection. In contrast, the mutant HGPRT was not protected from heat inactivation by PP-ribose-P concentrations as high as 1×10^{-2}M, a concentration which saturates the mutant enzyme under normal assay conditions. The failure of high PP-ribose-P concentrations to protect the mutant enzyme indicates that either PP-ribose-P was not significantly bound under the experimental conditions of high temperature and no purine substrate, or that binding of PP-ribose-P did not influence the rate of heat inactivation. Thus, the results of this heat inactivation study clearly indicate a major physical difference between the mutant and normal enzymes with respect to their behavior in the presence of PP-ribose-P.

In contrast to the kinetic differences between the normal and this mutant HGPRT, we have found that certain characteristics of the two enzymes are similar. First, the molecular weights of the two enzymes, measured in the absence of substrates by means of gel filtration on Sephadex G-100 and sucrose density gradient centri-

fugation, are both approximately 68,000. Second, despite the
disparate substrate kinetics of the two enzymes, the product, GMP,
inhibits the mutant HGPRT at least as well as the normal enzyme.

In addition to studying the physical and kinetic characteristics of this mutant HGPRT, we looked for the mutant enzyme in
erythrocytes from the patient's mother, a proven heterozygote for
the mutant enzyme (McDonald and Kelley, 1972). On the basis of
the Lyon (1961) hypothesis, individuals heterozygous for the Lesch-
Nyhan syndrome would be expected to have two populations of
erythrocytes in approximately equal proportions, one type with the
normal HGPRT and the other type exhibiting the mutant form of the
enzyme. In contrast to this prediction, previous studies by Nyhan
and co-workers utilizing another X-linked enzyme as a marker for
the HGPRT locus suggest that erythrocytes from such heterozygotes
consist largely of cells with the normal form of the enzyme (Nyhan
et al., 1970). In contrast to these indirect genetic studies, the
distinctive kinetic properties of this mutant HGPRT allow us to
measure it directly in the presence of the normal HGPRT without the
attendant difficulties of using linkage analysis.

The results of kinetic studies on hemolysates containing the
normal HGPRT, the mutant HGPRT, or an artificial equal mixture are

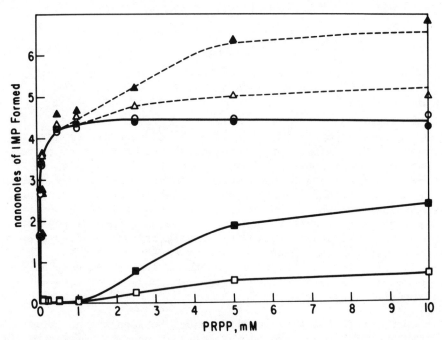

Fig. 6. Effect of variable PP-ribose-P concentration on rate of
IMP formation by normal HGPRT, mutant HGPRT, or an equal mixture
of both (From McDonald and Kelley, 1972).

both illustrated in Fig. 6. The solid line represents the effect of increasing PP-ribose-P concentration on IMP production by the normal HGPRT. The open circles represent activity at 0.1 mM hypoxanthine, while the closed circles represent activity at 1.0 mM hypoxanthine. Because of the low Km of the normal HGPRT for hypoxanthine, there is no difference in the two curves. The normal HGPRT reaches maximal activity at about 1.0 mM PP-ribose-P. The curve with the open squares represents the effect of increasing PP-ribose-P concentration on the mutant HGPRT assayed at 0.1 mM hypoxanthine, while the curve with the closed squares represents the activity of the mutant HGPRT assayed at 1.0 mM hypoxanthine. The higher hypoxanthine concentration results in increased activity because of the high Km of the mutant enzyme for hypoxanthine. In addition, the mutant HGPRT has minimal activity at PP-ribose-P concentrations below 2.5 mM, whereas substantial activity is observed at PP-ribose-P concentrations from 2.5 mM to 10.0 mM. The curves predicted from a mixture of normal and mutant hemolysates at low and high hypoxanthine are illustrated by the dashed lines. The actual values observed in such a mixture assayed at low and high hypoxanthine concentrations are given by the open and closed triangles, respectively. The activity of a mixture of the two enzymes is simply the sum of their independent activities, as seen from the excellent fit of the observed data points to the predicted curves. Thus, in a mixture of the two forms of HGPRT, the total activity at high PP-ribose-P and hypoxanthine concentration minus the activity at low PP-ribose-P concentration is due to the mutant HGPRT present in the mixture. Enzyme preparations from the patients mother showed no increase in activity at PP-ribose-P concentrations above 1.0 mM at either high or low hypoxanthine concentrations (Fig. 7). We were able to detect as low as 11% of the mutant HGPRT in artificial mixtures with hemolysates from the patient's mother. Accordingly, the mutant HGPRT appears to be absent from her erythrocytes. This heterozygote, like others previously examined, also had essentially normal HGPRT activity in hemolysates.

In summary, we have described a mutant form of the enzyme hypoxanthine-guanine phosphoribosyltransferase with increased Km's for both purine substrates guanine and hypoxanthine as well as PP-ribose-P and sigmoid kinetics with PP-ribose-P. The discovery of this mutant HGPRT provides proof that the enzyme defect in this individual is due to a structural gene mutation resulting in an altered enzyme protein. The studies of Kelley and Meade (1971), Rubin, et al. (1971), and Arnold, Meade and Kelley (1972) suggest, in fact, that this is the case for the majority if not all of the patients with the Lesch-Nyhan syndrome. The sigmoidicity of the PP-ribose-P rate curve observed with the mutant HGPRT may be due to positive homeotropic interaction of this substrate with the mutant enzyme. It is possible that this mutant HGPRT represents the production of allosterism from a normally nonallosteric enzyme by means of an apparent single structural gene mutation. Finally, we

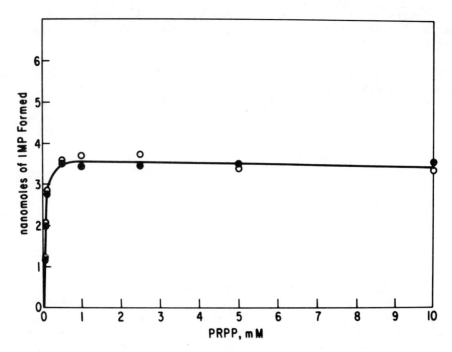

Fig. 7. Effect of variable PP-ribose-P concentration at a fixed high (●, 1.0 mM) or low (o, 0.1 mM) concentration of hypoxanthine on HGPRT activity from erythrocytes of a heterozygote for the mutant HGPRT (From McDonald and Kelley, 1972).

have demonstrated that the mutant enzyme is absent from erythrocytes obtained from a proven heterozygote. Thus, this mutant form of HGPRT provides an interesting model for examining the effect of a structural gene mutation upon kinetic properties of an enzyme and perhaps for elucidating the effect of an altered enzyme of purine metabolism upon cellular proliferation..

REFERENCES

Arnold, W. J., Meade, J. C. and Kelley, W. N. 1972. Hypoxanthine-guanine phosphoribosyltransferase: Characteristics of the mutant enzyme in erythrocytes from patients with the Lesch-Nyhan syndrome. J. Clin. Invest. 51: 1805-1812.

Kelley, W. N. and Meade, J. C. 1971. Studies on hypoxanthine-guanine phosphoribosyltransferase in fibroblasts from patients with the Lesch-Nyhan syndrome. J. Biol. Chem. 246: 2953-2958.

Lyon, M. F. 1961. Gene action in the X-chromosome of the mouse. Nature. 190: 372-373.

McDonald, J. A. and Kelley, W. N. 1971. Lesch-Nyhan syndrome: Altered kinetic properties of mutant enzyme. Science 171: 689-691.

McDonald, J. A. and Kelley, W. N. 1972. Lesch-Nyhan syndrome: Absence of the normal enzyme in erythrocytes of a heterozygote for both normal and mutant hypoxanthine-guanine phosphoribosyl-transferase. Biochem. Genetics 6: 21-26.

Nyhan, W. L., Bakay, B., Connor, J. D., Marks, J. F. and Keele, O. K. 1970. Hemizygous expression of glucose-6-phosphate dehydrogenase in erythrocytes of heterozygotes for the Lesch-Nyhan syndrome. Proc. Nat. Acad. Sci. 65: 214-218.

Rubin, C. S., Dancis, J., Yip, L. C., Nowinski, R. C. and Balis, M. E. 1971. Purification of IMP: Pyrophosphate phosphoribosyl-transferases, catalytically incompetent enzymes in Lesch-Nyhan disease. Proc. Nat. Acad. Sci. (U.S.A.) 68: 1461-1464.

Seegmiller, J. E., Rosenbloom, F. M. and Kelley, W. N. 1967. Enzyme defect associated with a sex-linked human neurological disorder and excessive purine synthesis. Science 155: 1682-1684.

HYPOXANTHINE-GUANINE PHOSPHORIBOSYLTRANSFERASE (HGPRT) DEFICIENCY:

IMMUNOLOGIC STUDIES ON THE MUTANT ENZYME

W. J. Arnold and W. N. Kelley

Department of Medicine, Duke University Medical Center

Durham, North Carolina 27710

The Lesch-Nyhan syndrome is a bizarre, X-linked disease characterized by spasticity, choreoathetosis, self-mutilation, mental and growth retardation as well as hyperuricemia and hyperuricaciduria (Lesch and Nyhan, 1964) which is due to a striking reduction of hypoxanthine-guanine phosphoribosyltransferase (HGPRT) activity in all tissues of affected individuals (Seegmiller, Rosenbloom and Kelley, 1967; Rosenbloom, et al., 1967). We have examined hemolysates from 14 patients and autopsy tissue from one patient with the Lesch-Nyhan syndrome for their content of HGPRT activity and immunologically detectable HGPRT enzyme protein by using a monospecific rabbit antiserum prepared against a homogeneous preparation of normal human HGPRT (Arnold, Meade and Kelley, 1972).

Table 1 shows HGPRT activity in fresh hemolysate from 5 patients with the Lesch-Nyhan syndrome and from 119 normal individuals. Hemolysates from 9 other patients with the Lesch-Nyhan syndrome which had been stored at -20° prior to assay showed no detectable HGPRT activity. Several other investigators have also noted low levels of HGPRT activity in erythrocytes (Mizuno, et al., 1970; Sorensen, 1970) and this finding is consistent with the low but readily detectable levels of HGPRT activity found in cultured fibroblasts from patients with the Lesch-Nyhan syndrome (Kelley and Meade, 1971).

Using the highly specific rabbit antiserum against normal human HGPRT we have been able to demonstrate material which crossreacts (CRM) with this antibody in hemolysate from all 14 patients and in autopsy tissue from an additional patient with the Lesch-Nyhan syndrome.

TABLE 1

HGPRT ACTIVITY IN HEMOLYSATES FROM PATIENTS WITH THE LESCH-NYHAN
SYNDROME

Patients	Specific Activity	
	Guanine	Hypoxanthine
	(nmoles/mg/hr)	
Normal	98±14	97±19
Lesch-Nyhan Syndrome		
W. E.	0.011	0.01
J. K.	0.023	0.01
D. C.	0.071	0.021
D. G.	0.038	0.01
E. S.	0.61	0.79
Nine Other Patients	< 0.001	< 0.01

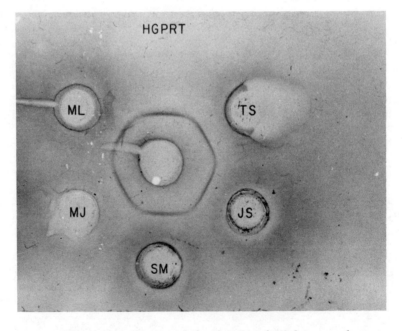

Fig. 1.   CRM in patients with the Lesch-Nyhan syndrome.

Several independent immunologic techniques were utilized to demonstrate CRM. Immunodiffusion using this monospecific antiserum revealed a single line of identity for normal HGPRT and for all 14 mutant hemolysates - five of which are shown in Fig. 1. Using this antiserum and semi-quantitative immunodiffusion, comparable levels of CRM were found in hemolysates from normal individuals and patients with the Lesch-Nyhan syndrome. Homogenates of liver and spleen from patient R. M. also exhibited a single line of identity with normal human HGPRT as illustrated in Fig. 2. Table 2 compares the amount of CRM in tissues from patient R. M. with the amount of assayable HGPRT activity. As will be noted, despite low levels of HGPRT activity throughout, CRM is always detectable and is highest in amount in brain where the highest level of HGPRT activity is normally found.

The highly specific HGPRT antiserum could inhibit the catalytic activity of the normal enzyme (Fig. 3). By adding a constant amount of antiserum to successive dilutions of antigen, near maximal inhibition of enzyme activity was achieved at a 1:32 dilution of hemolysate.   This antiserum was also shown to inhibit the catalytic activity of the mutant HGPRT from patient E.S.   Similar patterns of inhibition were obtained for both the normal and this specific mutant form of HGPRT.

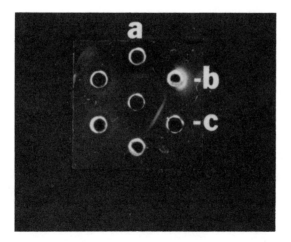

Fig. 2. CRM in tissues from subject R. M. with the Lesch-Nyhan syndrome.   A.   Liver,   B.   Spleen,   and   C.   Normal hemolysate.

TABLE 2

COMPARISON OF IMMUNOREACTIVE HGPRT PROTEIN (CRM) AND ASSAYABLE HGPRT
ACTIVITY IN TISSUES FROM PATIENT R.M. WITH THE LESCH NYHAN SYNDROME

Tissue	CRM	Specific Activity
	(μg HGPRT/gm tissue)	(nmoles/mg protein/hr)
Brain	456	1.5
Spleen	280	1.0
Liver	228	0.61
Testis	116	0.43

Incubation of antiserum with hemolysate from each of the fourteen patients studied with the Lesch-Nyhan syndrome led to removal of HGPRT antibody as demonstrated by reversal of the antibody-mediated inhibition of the normal enzyme (Table 3). In this case, unabsorbed antiserum produced 42% inhibition of normal HGPRT and this inhibition was essentially completely reversed when the antiserum was pre-incubated with mutant hemolysate before being added to normal hemolysate. Thus, in each of the 14 hemolysates studied, material was present which was capable of reacting with the HGPRT antibody such that it was no longer free to inhibit the normal enzyme. This technique was also used by Rubin, et al (1971) to demonstrate CRM in 4 additional patients with the Lesch-Nyhan syndrome.

These studies demonstrate that hemolysates from 14 patients with the Lesch-Nyhan syndrome contain normal amounts of a catalytically-incompetent HGPRT protein. CRM is also present in other tissues. Thus, these studies provide additional evidence that in most if not all instances a mutation on the structural gene coding for HGPRT is responsible for the development of the Lesch-Nyhan syndrome.

The low level of HGPRT activity found in hemolysate from 5 patients with the Lesch-Nyhan syndrome was further evaluated by density fractionation of fresh, intact circulating erythrocytes. Each fraction was assayed for HGPRT activity and the amount of immunologically-detectable HGPRT protein present in the youngest

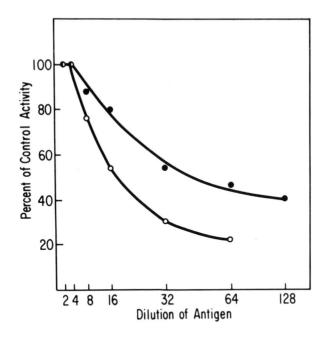

Fig. 3. Antibody-mediated inhibition of normal (●-●) and mutant (○-○) HGPRT. Antigen represents HGPRT in hemolysate. (From Arnold, Meade and Kelley, 1972).

TABLE 3

REMOVAL OF HGPRT ANTIBODY BY INCUBATION WITH LESCH-NYHAN HEMOLYSATES

	Inhibition
	(%)
Before Incubation	42
After Incubation (n = 14 patients)	3.3±3.9
	(Range 0-10)

and oldest cell fractions was determined.   In Fig. 4 it can be
seen that the HGPRT activity present in normal erythrocytes is from
100 to 1000 times greater than that found in patients with the
Lesch-Nyhan syndrome.   Also, the HGPRT activity does not decrease
in normal erythrocytes with increasing age indicating that the activity of normal erythrocyte HGPRT is stable in vivo.   In contrast,
the mutant HGPRT activity from at least three of the 5 patients
with the Lesch-Nyhan syndrome decreased in the older erythrocytes
suggesting a relative instability of the mutant enzyme activity in
comparison to the normal.   In the 4th patient (D.G.) the level of
HGPRT activity may also have declined in the oldest cells while in

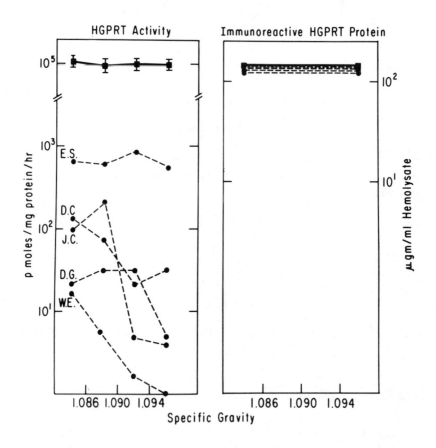

Fig. 4.   Comparison of HGPRT activity and immunoreactive HGPRT protein in erythrocytes separated by their density. ■–■ , normal; ●–●
Lesch-Nyhan syndrome (From Kelley, and Arnold, Fed. Proc., 1973).

the 5th patient (E.S.) the HGPRT activity appeared stable. The
right hand panel of Fig. 4 illustrates the results of CRM determinations in the youngest and oldest erythrocytes from these patients
and shows that despite a striking reduction in HGPRT activity, the
amount of HGPRT enzyme protein remains unchanged. This finding is
perhaps comparable to previous studies in cultured fibroblasts from
patients with the Lesch-Nyhan syndrome in which the HGPRT enzyme from
7 of 8 mutant strains studied (including cell strains derived from
J.K. and W.E.) was found to be abnormally labile to heat treatment.
However, unlike the loss of HGPRT activity in vivo, the decrease in
HGPRT activity seen after heat treatment at 82° is accompanied by a
corresponding decrease in the amount of CRM (Fig. 5) suggesting that
the two situations are not truly comparable.

These data not only confirm the heterogeneity of the mutations
on the structural gene coding for HGPRT which was first suggested
by studies of cultured fibroblasts from affected individuals (Kelley
and Meade, 1971) but also suggest an additional manifestation of
the mutation in some patients, a reduced half-life of enzyme activity
inccirculating erythrocytes.

In summary, the data presented in this report indicate that:

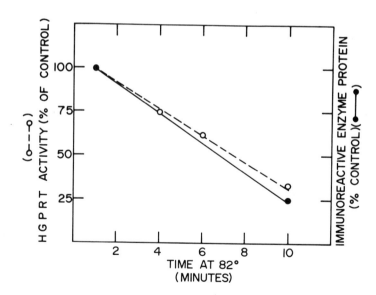

Fig. 5. Relationship of HGPRT activity and immunoreactive HGPRT
enzyme protein during heat treatment (82°).

1) In most, if not all instances of the Lesch-Nyhan syndrome, a mutation(s) on the structural gene coding for HGPRT results in the synthesis of a normal amount of a catalytically-defective HGPRT protein. This immunologically-detectable yet virtually inactive HGPRT protein is present not only in hemolysate but also in brain, spleen, liver and testis with levels approximately that expected in a normal individual. 2) As an additional manifestation of the mutation, HGPRT activity from at least 3 of 5 patients with the Lesch-Nyhan syndrome had a shortened half-life in vivo which was not associated with an accelerated degradation of HGPRT enzyme protein.

Finally, therapy directed at activation or stabilization of this enzyme activity in vivo is feasible and could be assessed serially by direct assay of HGPRT activity in circulating erythrocytes separated by their density.

REFERENCES

Arnold, W. J., Meade, J. C. and Kelley, W. N. 1972. Hypoxanthine-guanine phosphoribosyltransferase: Characteristics of the mutant enzyme in erythrocytes from patients with the Lesch-Nyhan syndrome. J. Clin. Invest. $\underline{51}$: 1805-1812.

Kelley, W. N. and Meade, J. C. 1971. Studies on hypoxanthine-guanine phosphoribosyltransferase in fibroblasts from patients with the Lesch-Nyhan syndrome. Evidence for genetic heterogeneity. J. Biol. Chem. $\underline{246}$: 2953-2958.

Kelley, W.N. and Arnold, W.J. Human hypoxanthine-guanine phosphoribosyltransferase: Studies on the normal and mutant forms of the enzyme. Fed. Proc. (in press).

Lesch, M. and Nyhan, W. L. 1964. A familial disorder of uric acid metabolism and central nervous system function. Amer. J. Med. $\underline{36}$: 561-570.

Mizuno, T., Segawa, M., Kurumada, T., Maruyama, H. and Onisawa, J. 1970. Clinical and therapeutic aspects of the Lesch-Nyhan syndrome in Japanese children. Neuropaediatrie. $\underline{2}$: 38-52.

Rosenbloom, F. M., Kelley, W. N., Miller, J. M., Henderson, J. F. and Seegmiller, J. E. 1967. Inherited disorder of purine metabolism: Correlation between central nervous system dysfunction and biochemical defects. J.A.M.A. $\underline{202}$: 175-177.

Rubin, C. S., Dancis, J., Yip, L. C., Nawinski, R. C. and Balis, M. E. 1971. Purification of IMP: Pyrophosphate phosphoribosyltransferases, catalytically incompetent enzymes in

Lesch-Nyhan disease.   Proc. Nat. Acad. Sci.(U.S.A.)   68: 1461-1464.

Seegmiller, J. E., Rosenbloom, F. M. and Kelley, W. N.   1967.   An enzyme defect associated with a sex-linked human neurological disorder and excessive purine synthesis.   Science 155: 1682-1684.

Sorensen, L.   1970.   Mechanism of excessive purine biosynthesis in hypoxanthine-guanine phosphoribosyltransferase deficiency. J. Clin. Invest. 49: 968-978.

IMMUNOLOGICAL STUDIES OF HYPOXANTHINE-GUANINE

PHOSPHORIBOSYLTRANSFERASE IN LESCH-NYHAN SYNDROME

M. M. Müller and H. Stemberger

Departments of Medical Chemistry and Hygiene

University of Vienna, Austria

The clinical picture of the Lesch-Nyhan syndrome shows besides massive hyperuricemia mental retardation and spasticity (later choreoathetosis and self mutilation in the behavior) (1). The disturbance of purine metabolism is based on the deficiency of hypoxanthine-guanine phosphoribosyltransferase activity (HG-PRT) (2).

The purpose of the present study is to ascertain whether the enzymatic defect is based on
(1) decreased synthesis and/or
(2) structural changes of the enzyme protein.

MATERIALS AND METHODS

Patients: For our investigations we used the hemolysates of two children with classical clinical symptomes of the Lesch-Nyhan syndrome. The erythrocytes of a metabolically healthy male were used for control.

Measurements of enzyme activities: The enzyme activities of HG-PRT and adenine phosphoribosyl-transferase (A-PRT) of the washed and hemolysed erythrocytes of the test persons were determined according to the radiochemical method of Kelley (3). As substrates we used hypoxanthine and adenine respectively. The hemolysates were incubated together with the substrates for 20 minutes at 37°C. For

measuring thermostability of the enzymes the hemolysates
were incubated with buffer for 20 minutes at different
temperatures. The test tubes were cooled, substrates
added and the enzymo activities were determined.

Production of antienzyme serum: HG-PRT was
isolated from 500 ml blood of a male donor with normal
HG-PRT activity in the red cells, as recently
reported (4). With this enzyme preparation rabbits
were immunised. The resulting antiserum contained
antibodies against HG-PRT and also against a protein
which was not more precisely defined, but which was
not identical with a serum protein. Further
purification of the antiserum was done by means of
immunelectrophoresis: The enzymepreparation was run
against the antiserum. The precipitin lines
corresponding to the enzyme were cut out of the agargel.
Now rabbits were now immunised with these immuno-
complexes. The resulting antiserum was used for
our further investigations.

Antibody consumption tests: For the quantitative
determination of the enzyme protein in the hemolysates,
we incubated 20 µl hemolysate of both patients and
control (previously adjusted to the same protein level)
together with 200 µl antiserum and 50 µl buffer at
37°C for one hour. Following this the mixtures were
cooled in an icebath for 20 minutes and the immuno-
precipitates were removed by centrifugation at
5.000 x g for 20 minutes. The amount of the remaining
antibodies in the supernatant (50 µl) was measured
by direct enzyme inhibition using 20 µl of a normal
hemolysate.

Enzyme inhibition tests: The direct inhibition of
HG-PRT was achieved by the addition of increasing
amounts (50 - 200 µl) of antiserum (4,75 mg protein
per ml) to the enzyme assay. For control we used
normal rabbit serum adjusted to the same protein
concentration as the antiserum. Substrates, antiserum
and 20 µl hemolysate from the control person and that
of the patient H. H. (again both had the same
protein content: 10 mg per ml) were incubated for
20 minutes at 37°C.

Miscellaneous techniques: Ouchterlony tests
were done on immunoplates of Hyland and protein
determinations according to Sols (5).

RESULTS AND DISCUSSION

Enzyme Activities

The results of the determinations of the enzyme activities in the erythrocytes are summerized in table 1.

Patient	HG-PRT nM IMP/mg protein/hr	A-PRT nM AMP/mg Protein/hr
normal (8)	$78,60 \pm 11,51$	$15,36 \pm 2,44$
H. H.	3,84	19,80
O. K.	0,68	26,79

Table 1. HG-PRT and A-PRT activities in hemolysates from patients with Lesch-Nyhan syndrome.

As it had allready been described (6), our patients with Lesch-Nyhan syndrome showed extremely low HG-PRT activities in the erythrocytes. The enzyme activity of patient H. H. was 4,9%, the activity of patient O. K. was 0,9% of the normal value. It is remarkable that the activities of A-PRT were elevated in both patients: the increase was 29% in the patient H. H.'s erythrocytes, 74% in the patient O. K.'s.

Ouchterlony tests

In the agar gel double diffusion test a band with two precipitin lines appeared when normal hemolysate reacted with antiserum (Fig. 1). Two precipitin lines were formed when we used the patients' hemolysates in the Ouchterlony test (Fig. 1). Inspite of the deficiency of enzyme activities in the erythrocytes of our patients these precipitin lines showed a complete identity with the lines of the hemolysate from the healthy control person.

The appearance of two precipitin lines in the Ouchterlony test could be explained as follows:
(1) Inspite of our immunisation technique the antiserum was not monospecific, it reacted with HG-PRT and with another protein.

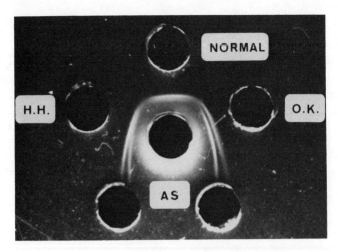

Figure 1. Ouchterlony tests of normal hemolysate
and hemolysates of patients with Lesch-Nyhan syndrome.

(2) The antiserum was monospecific, but it reacted
    with two enzyme proteins. There could exist a
    splitting of the precipitin line, as described for
    paraproteins in immunoelectrophoresis.
Accirding to our opinion there exist two different
enzyme proteins, an active and an inactive one, both
capable of reacting with the antiserum. It is
remarkable that the splitting of the precipitin line
occurs particularly using patient O. K.'s hemolysate
which was nearly without enzyme activity.

Antibody Consumption

Antibody consumption tests showed equal amounts
of antibodies in the supernatant of both normal and
Lesch-Nyhan hemolysates (table 2):

Patient	Antiserum added µl	Percent of initial antiserum activity
normal	200	26
H. H.	200	30
O. K.	200	25

Table 2. Antibody consumption tests.

In the Ouchterlony test with dilutions of hemolysates at constant antiserum level precipitin lines occured down to 1:16 antigen dilution.

The identical antibody consumption and the results with serial dilutions of hemolysate in the Ouchterlony tests suggest that the amount of enzyme protein in Lesch-Nyhan patients is just as high as that synthesized in normal persons.

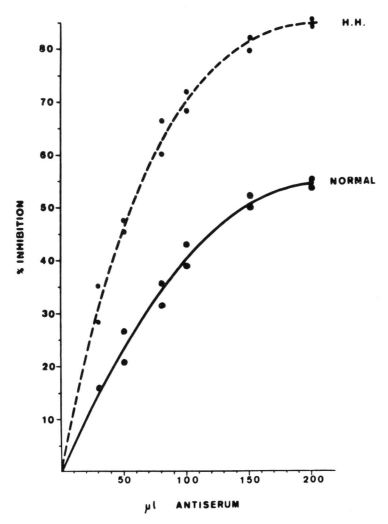

Figure 2. Direct enzyme inhibition of HG-PRT by antiserum using the hemolysate of a normal person and that of the Lesch-Nyhan patient H. H.

Enzyme Inhibition

The enzyme inhibition tests demonstrated the biological specificity of the antiserum. By adding equal amounts of normal hemolysate and hemolysate of the patient H. H. to increasing amounts of antiserum, we could obtain a dose dependent inhibition of enzyme activity in both cases. However the Lesch-Nyhan enzyme was more susceptible to the inhibition by antiserum (Fig. 2).

Thermostability

Besides, we compared the temperature sensitivity of the normal enzyme with that of the enzyme of patient H. H.. The hemolysates (adjusted to the same protein level) were previously incubated at different temperatures for 20 minutes without substrates. Then the enzyme activities were determined. The normal enzyme showed an obvious drop in activity at 75°C, while the patient H. H.'s enzyme showed a comparable drop in its activity allready at 56°C. (Fig. 3)

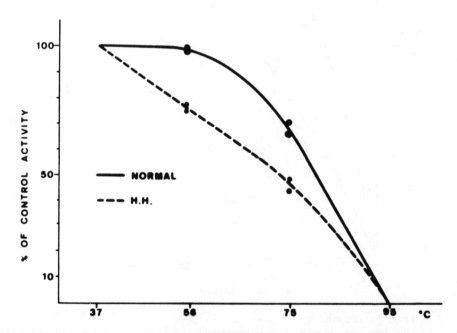

Figure 3. Temperature sensitivity of HG-PRT in hemolysates (normal person and Lesch-Nyhan patient H.H.)

This demonstrates not only that HG-PRT of a patient with Lesch-Nyhan syndrome can be more strongly inhibited by antibodies but it also exhibits a stronger heat lability.

CONCLUSION

The introductory questions could be answered being based on our studies in this manner:
(1) In a patient with Lesch-Nyhan syndrome an almost inactive enzyme protein is synthesized in normal amounts; this enzyme is immunologically identical with the normal enzyme protein.
(2) The enzyme protein in Lesch-Nyhan patients is more heterogeneous, as demonstrated by the Ouchterlony test, and more susceptible to antibody inhibition and thermal inactivation.

SUMMARY

(1) In two cases of Lesch_Nyhan syndrome with complete and partial deficiency of HG-PRT activity the presence of enzyme protein could be demonstrated using an antiserum prepared against the enzyme from normal erythrocytes. The immunoreactive proteins in hemolysates of both normal and sick patients showed a complete identity.
(2) Immunodiffusion and antibody consumption tests using normal and sick patients' hemolysates revealed equal amounts of enzyme protein.
(3) Increasing amounts of antiserum inhibited the enzyme activity of both normal erythrocytes and of the erythrocytes of the Lesch-Nyhan patient with the partial enzyme deficiency.
(4) The enzyme of the Lesch-Nyhan patient with the partial enzyme deficiency was more susceptible to thermal inactivation.

REFERENCES

(1) Lesch M., Nyhan W.L.: Amer. J. Med. 36, 561 (1964)
(2) Seegmiller J.E., Rosenbloom F.M., Kelley W.N.: Science 155, 1682 (1967)
(3) Kelley W,N., Rosenbloom F.M., Henderson J.F., Seegmiller J.E.: Proc. Nat. Acad. Sci, USA 57, 1735 (1967)

(4)   Müller M.M., Dobrovits H., Stemberger H.:
           Z. Klin. Chem. Klin. Biochem. 10, 535 (1972)
(5)   Sols A.: Nature (London) 160, 89 (1947)
(6)   Rubin C.S., Balis M.E., Piomelli S., Berman P.H.,
       Dancis J.: J. Lab. Clin. Med. 74, 732 (1969)

UNSTABLE HPRTase IN SUBJECTS WITH ABNORMAL URINARY OXYPURINE EXCRETION

M.E. Balis, L.C. Yip, T.F. Yü, A.B. Gutman*, R. Cox
and J. Dancis
Sloan-Kettering Memorial Cancer Center
Mt. Sinai School of Medicine of the City University of
New York
New York University School of Medicine, New York, N.Y.

Several genetic defects have been recognized which have presented different levels of residual IMP pyrophosphorylase (HPRTase) activity and concomitant modifications of the clinical manifestations. The most extreme is Lesch-Nyhan disease with little or no detectable enzyme in the red cells and severe neurological damage.[1] If the enzyme deficiency is less complete, there may be only overproduction of uric acid and gout.

Evidence for structural enzyme defect

Antibody was prepared against purified HPRTase and used to measure enzyme protein in lysates of erythrocytes from five Lesch-Nyhan subjects.[2] In all instances essentially normal amounts of an immunologically identifiable but catalytically incompetent enzyme was demonstrated. A rat and rabbit antibody to the enzyme from human erythrocytes was used. Both antibodies neutralized the catalytic property of the enzyme. Lysates from five patients with Lesch-Nyhan disease reacted with these antisera and blocked the neutralization of normal enzyme. The amount of immunologically active material was determined from the volume of lysate required to achieve maximal neutralization of a given volume of antiserum. The amount was also evaluated by double diffusion studies in which anti-HPRTase serum was incorporated into an agar layer and fixed volumes of red cell lysates were permitted to

---

*Deceased

diffuse through the agar. From the size of the precipitin area
it is possible to calculate the amount of cross-reacting material
present in the erythrocytes of the patients. Despite the fact
that the amount of immunologically reacting material appeared to
be quantitatively identical, neutralization assays of erythrocyte
lysates from Lesch-Nyhan patients revealed differences, although
total antigen was the same by double diffusion (Fig. 1).

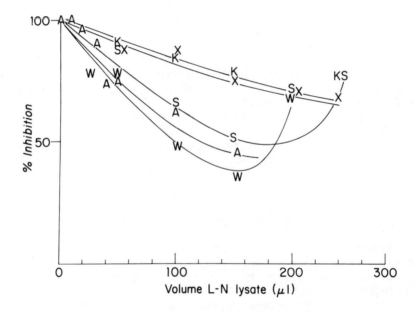

Figure 1. IMMUNOASSAY OF LESCH-NYHAN HEMOLYSATES

    Rat anti-human IMP pyrophosphorylase globulin (AB) was ob-
tained by adding solid ammonium sulfate to immune serum (50%
saturation). The reaction was carried out by adding to DEAE-
cellulose treated blood to appropriate amounts of DEAE-cellulose
treated Lesch-Nyhan blood and potassium phosphate buffer 0.01 M,
pH 7.0, to a final volume of 250 µl. The mixture was incubated
at 4° for 16 hours, centrifuged at 2000 RPM for 30 minutes and
50 µl of the supernatant was assayed for IMP pyrophosphorylase
activity.

We have also studied several individual members of families carrying a genetic defect resulting in partial HPRTase deficiency and hyperuricemia.[3] Neutralization tests with lysates from these variants have also demonstrated presence of enzyme antigen. The neutralization curves differ from the normal and from each other (Fig. 2), suggesting different modifications in protein structure among the variants. There were also differences in enzyme stability measured in vitro.

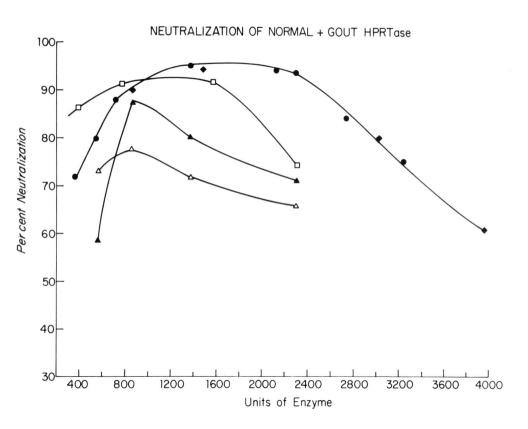

Figure 2.   IMMUNOASSAY OF HPRTase OF GOUTY VARIANTS

Varying volumes of 1:1 mixtures of lysates of normal and gouty erythrocytes were mixed with a fixed amount of rat anti-HPRTase and globulin. Phosphate buffer was added to final volume of 200 μl. The mixture was incubated at 4° for 16 hours, centrifuged and the supernatant assayed for HPRTase activity.
0 is normal. ☐ is gouty over-excreter with normal HPRTase. Others are over-excreters with low HPRTase.

Relative enzyme activity in erythrocytes and leukocytes

     Two cousins, both males, presented with no detectable HPRTase activity in the erythrocytes but without the symptomatology of Lesch-Nyhan disease. However, leukocyte lysates retained about 15% of normal activity (Table 1).

TABLE 1

HPRTase ACTIVITY

	RBC (lysate + PRPP)	WBC (lysate + PRPP)	WBC (intact)
Ro family			
Mother - Cl	23	35	.40
Daughter - Ju	126	80	---
Son - Ho*	<0.1	11	.36
McN family			
Mother - Ed	11	23	.46
Daughter - An	122	80	.35
Son - Be	97	76	.48
Son - Ro*	<0.1	7.5	.47
Normal	107±26	67±21	.4±1

HPRT activity in mμ moles product/mg protein/hour.
HPRT assays were performed on RBC and WBC lysates with excess PRPP (2.5×10^{-2}M). Intact leukocytes were incubated with hypoxanthine without added PRPP.[4]

*Hemizygotes

The stability of the enzyme was measured by storing white cells
and white cell lysates at 4°C and noting the decay rate. Enzyme
activity in normal subjects actually increased during 24-48 hours
while it fell rapidly in the patients (Table 2). The results with
APRTase were normal (Table 2).

TABLE 2

STABILITY OF ENZYME IN LEUKOCYTES OF Ro FAMILY

Age of Extracted Cells	HoRo		C1Ro (Mother)		Normal	
	APRT	HPRT	APRT	HPRT	APRT	HPRT
Fresh	3800	2000	3000	6300	5600	11000
24 hour	5100	610	3600	4300	6600	18000
96 hour	5800	50	4200	3400	7200	19000

Leukocyte lysates adjusted to equal protein content were maintained at 4° in buffer. At indicated times aliquots were removed
and enzyme activity determined. The size of the aliquot was constant for each subject. The assay mixture contained 10 μ liters
1M tris buffer, pH 8.0, 50 μ liters 0.02 M $MgCl_2$ and 50 μ liters
10^{-3}M PRPP (mixed just before assay; 110 μ liters of the mixture
used per tube), 50 μ liters of 1:50 lysate. The reaction was
started by adding 8^{-14}C-hypoxanthine, or 8^{-14}C-adenine, 5 x 10^{-4}M.

Results in counts per minute per assay.

The enzyme activity showed a slight increase on storage as has
been reported previously. Similar instability was demonstrated
in other variants (Table 3). These patients had been studied
previously and shown to have low values of HPRTase. The patient R
had much activity than the others and his enzyme was more stable.
Subjects Bo, D'A and F were all members of the same family. All
had essentially equal levels of enzyme activity and roughly equal
rates of inactivation on storage.

TABLE 3

STABILITY OF HPRTase IN LEUKOCYTE LYSATES

OF GOUT PATIENTS

	Red Cell	Fresh Extract	24 hour	48 hour
Controls	3290	5495	4892	4901
I	4	17	0	0
R	580	800	----	260
Bo	5	58	39	6
D'A	12	50	42	7
F	8	39	19	3

Assay was performed as described in Table 2. Activity is
given as counts/minutes IMP formed/mg protein.

A reasonable explanation of the disparate activity in erythrocytes and leukocytes is that the anucleated erythrocyte is unable to synthesize enzyme and rapidly loses the activity of the unstable enzyme, while the leukocyte continues to synthesize enough enzyme to maintain a reduced but easily measurable level. Intermediate results in the mothers indicate mosaicism in the circulating blood cells, differing in this respect from the heterozygote in L.N. disease and conforming to previous speculations concerning embryonic selection of stem cells.[5]

Enzyme activity in intact vs. lysed leukocytes

HPRTase assays are usually performed with lysates and saturating concentrations of PRPP. A more valid indication of in vivo performance might be obtained by incubating intact cells with physiological concentrations of Hx and relying on endogenous PRPP. With this technique, the two cousins described above had normal enzyme activity. Appropriate tests demonstrated that PRPP may be rate limiting even at reduced levels of enzyme activity.[6]

Summary:

The enzyme defect in L.N. disease and several variants have been studied immunologically with results indicating a structural alteration in the enzyme. Enzyme function is reduced by modifying the competence of the enzyme and its stability. Disparate reductions in enzyme activity in erythrocyte and leukocyte lysates, the latter being less severely affected, may reflect instability of the enzyme. Measurement of enzyme levels in the erythrocyte in such cases may provide an inaccurate evaluation of the severity of the defect in the patient. Enzyme assays with the intact leukocyte may reflect in vivo function more accurately than the customary assay with lysates using high levels of substrate and saturating amounts of PRPP.

ACKNOWLEDGEMENT

This research was supported in part by U.S. Public Health Service Research Grants CA 08748 and HD 04526 and by the National Foundation - March of Dimes CRBS-232

REFERENCES

1. Lesch, M. and Nyhan, W.L. (1964) Amer. J. Med. 36, 561-570.
2. Rubin, Charles S., Dancis, J., Yip, L.C.. Nowinski, R.C. and
    Balis, M.E. (1971) Proc. Nat. Acad. Sci. 68, 1461-1464.
3. Yu, T-F., Balis, M.E., Krenitsy, T.A., Dancis, J.,
    Silvers, D.N., Elion, G.B. and Gutman, A.B. (1972) Ann. Intern.
    Med. 76, 255-264.
4. Smith, J.L., Omura, G.A., Krakoff, I.H. and Balis, M.E. (1971)
    Proc. Soc. Exp. Biol. and Med. 136, 1299-1303.
5. Dancis, J., Berman, P.H., Jansen, V. and Balis, M.E. (1968)
    Life Sciences 7, Part II, 587-91.
6. Dancis, J., Yip, L.C., Cox, R.P., Piomelli, S. and Balis, M.E.
    J. Clin. Invest. (In Press).

HYPOXANTHINE-GUANINE PHOSPHORIBOSYLTRANSFERASE (HGPRT) DEFICIENCY:

EFFECT OF DIETARY PURINES ON ENZYME ACTIVITY

W. J. Arnold and W. N. Kelley

Department of Medicine, Duke University Medical Center

Durham, North Carolina   27710

The X-linked, virtually complete deficiency of hypoxanthine-guanine phosphoribosyltransferase (HGPRT) activity results in a bizarre neurologic disorder, the Lesch-Nyhan syndrome (Seegmiller, Rosenbloom and Kelley, 1967). In initial reports HGPRT activity was noted as undetectable in erythrocytes from patients with the Lesch-Nyhan syndrome. However, more recently several investigators have noted low but detectable levels of HGPRT activity in erythrocytes from these patients (Mizuno, et al., 1970; Sorenson, 1970). In addition, it is now known that despite this striking reduction in HGPRT activity there is a normal amount of immunologically detectable HGPRT protein present (Rubin, et al., 1971; Arnold, Meade and Kelley 1972). We subsequently found that a given patient would also exhibit a wide variability in HGPRT activity even when assay conditions were standardized. This suggested that environmental factors might be responsible for changes in enzyme actiivty and that one potential approach to therapy of patients with this disease would be activation of the structurally abnormal HGPRT protein. The present report describes the influence of alterations in dietary purine content on the erythrocyte HGPRT activity from three patients with the Lesch-Nyhan syndrome.

All three patients were hospitalized on the Clinical Research Unit of Duke University Medical Center and received isocaloric diets which, as specified later, contained a normal amount of dietary purines, were essentially free of purines or contained the amount of adenine (10 mg/Kg/day in four divided doses) roughly equivalent to the normal daily intake. All three patients received allopurinol (150-300 mg/day) orally throughout the entire period of study.

Figure 1 illustrates the effect of the presence or absence of

203

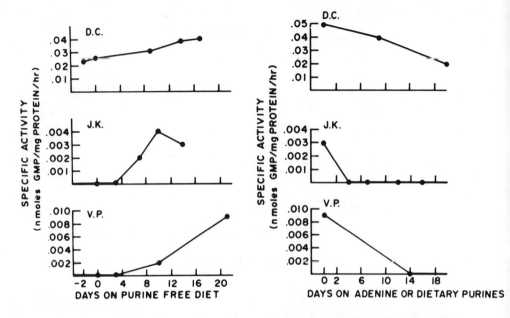

Fig. 1.   The effect of the dietary purines or adenine on erythrocyte HGPRT activity.   The panel on the left illustrates the effect
of purine restriction on erythrocyte HGPRT activity while the panel
on the right illustrates the effect of normal dietary purine content
either by the administration of adenine (D.C. and J.K.) or by
reinstitution of a diet containing a normal amount of purines (V.P.)
on HGPRT activity.   (From Arnold and Kelley, 1973).

dietary purines on the level of HGPRT activity present in hemolysate from three patients.   During the period of dietary purine
restriction, all three patients demonstrated an increase in HGPRT
activity.   In two of the three patients the activity was initially
undetectable and rose at least 3 and 9 fold, respectively while
in the third patient an increase of 54% above control values was
observed.   The right hand panel demonstrates the decrease in HGPRT
activity seen when adenine was added to a purine free diet (patients
D.C. and J.K.) or when normal dietary purine intake was resumed
(patient V.P.).   In V.P. and J.K. erythrocyte HGPRT activity again
became undetectable while in the third patient (D.C.) the HGPRT
activity decreased by 58%. The activity of ODC and OPRT was unchanged
throughout the entire study period.

      The results of density fractionation of intact circulating
erythrocytes from patient D.C. before and after 9 and 18 days of
adenine administration are shown in Figure 2.   As can be seen the

decrease in HGPRT activity occurred to approximately the same extent
in both the youngest (specific gravity <1.086) and oldest (specific
gravity >1.096) cell fractions without a change in the apparent
half-life of HGPRT activity.

We have considered the possibility that these results were due
to assay variability or artifact. Several additional experiments
were performed which suggest that this is not the case: 1) Identification of the product of the reaction mixture (C^{14}-IMP or C^{14}-GMP)
was done in 4 different chromatography systems.   2) The standard
error of the mean observed from 10 different determinations of HGPRT
activity in a single sample of hemolysate from D.C. and V.P. was 2.9%
and 5.2% of the mean activity respectively. Also the overall trend
of HGPRT activity in all three patients in response to manipulation
of dietary purines was consistent.

We have examined erythrocyte hemolysate from all three patients

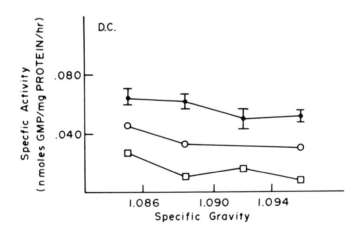

Fig. 2. The effect of adenine administration (10 mg/kg/day) on
HGPRT activity in density-fractionated erythrocytes from patient
D.C. Range of two separate determinations during the control
period (vertical bars); 9 days after beginning adenine (O—O); 14
days after beginning adenine (□—□). (From Arnold and Kelley, 1973).

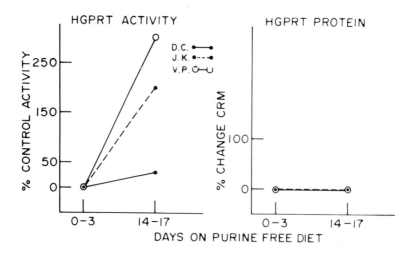

Fig. 3. Effect of purine free diet on HGPRT activity and immunologically-detectable HGPRT protein (CRM).

for their content of immunoreactive HGPRT protein using quantitative
immunodiffusion and a monospecific rabbit antisera prepared against
a homogeneous preparation of normal human HGPRT. This technique
was standardized by using known quantities of pure human HGPRT.
Changes in concentration of 13 μg/ml could be detected (Arnold,
Meade and Kelley, 1972). In Figure 3 it can be seen that despite
the increase in HGPRT activity in all 3 patients while on purine
restriction there was no change in the amount of immunoreactive
HGPRT protein. These data suggest that the changes in HGPRT activity
observed during purine restriction are due to a post-translational
alteration of the catalytically incompetent HGPRT protein.

To elucidate the nature of this post-translational alteration
we have examined the effect of adenine and other compounds on the
stability of HGPRT activity from patient D.C. in vitro. Since
normal human HGPRT is stabilized to heat treatment by 5-phosphori-
bosyl-1-pyrophosphate or PP-ribose-P it is possible that the
decreased HGPRT activity seen during adenine administration is due to
depletion of erythrocyte PP-ribose-P. Figure 4 illustrates the effect
of increasing concentrations of PP-ribose-P on the heat inactivation of HGPRT activity from patient D.C. Concentrations of
PP-ribose-P twice that found in this patient (5×10^{-5}M) did not
offer significant protection from heat inactivation. Preincubation
of undialyzed hemolysate from patient D.C. with either 0.1 mM
adenine or 0.1 mM aminoimidazole carboxamide (AIC) appeared to

decrease the heat stability of HGPRT after 4 minutes at 80° although
this effect was not pronounced (Figure 5).

These studies demonstrate that exogenous influences in the
form of dietary purines can alter the phenotypic expression of the
basic genetic defect responsible for the Lesch-Nyhan syndrome in
some patients with this disorder, namely the level of HGPRT
activity. At present the mechanism of this alteration in HGPRT
activity is obscure. We are unable based on our present data to
relate the decrease in HGPRT activity to the depletion of intracellular PP-ribose-P which is a known effect of adenine administration (Schulman, et al., 1971). The changes of HGPRT activity
observed in this study are also difficult to relate to the apparent
activation of mutant HGPRT observed in vitro by Bakay and Nyhan
(1972) since the activation reported by these investigators required
the presence of normal HGPRT protein.

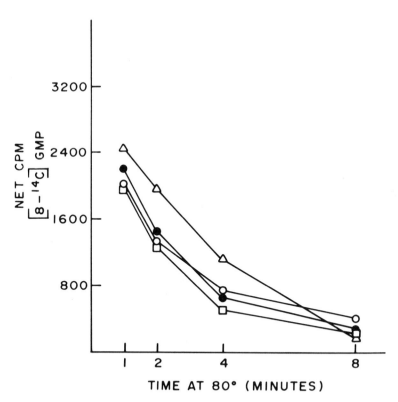

Fig. 4.   Effect of PP-ribose-P on inactivation of HGPRT from patient
D.C. at 80°.   Control,●—●; $5\text{X}10^{-6}$M PP-ribose-P,○—○; $5\text{X}10^{-5}$M PP-ribose-P,□—□; $5\text{X}10^{-4}$M PP-ribose-P,●—● .

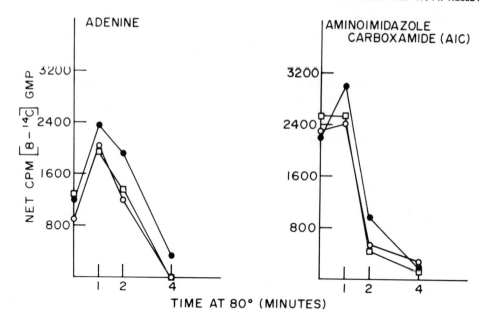

Fig. 5. Effect of preincubation with adenine and amino-imidazole-carboxamide on heat stability of HGPRT activity in erythrocytes from a patient with the Lesch-Nyhan syndrome. Control,●—●; 0.1 mM adenine or AIC,○—○; 0.01 mM adenine or AIC,□—□.

We have recently treated one additional patient with a purine free diet. Subject E.S., who has been shown to have an altered Km for PP-ribose-P, hypoxanthine and guanine, had no demonstrable change in enzyme activity over a 3 week period. Thus it is clear that this effect cannot be demonstrated in every patient with a deficiency of HGPRT.

These observations provide one explanation for the variability of HGPRT activity observed with time in a given patient with the classical Lesch-Nyhan syndrome. In addition, although we did not observe any striking alterations in the clinical course of any patients during purine restriction it is possible that the beneficial effects may become evident after more prolonged dietary restriction. We have now treated one patient, D.C., with dietary purine restriction for 6 months as an outpatient. There was no change in behaviour nor was there a further increase in enzyme activity. However, we have no independent estimate that the diet he received during this period was in fact purine free.

REFERENCES

Arnold, W. J., Meade, J. C., and Kelley, W. N. 1972. Hypoxanthine-guanine phosphoribosyltransferase: Characteristics of the mutant enzyme in erythrocytes from patients with the Lesch-Nyhan syndrome. J. Clin. Invest. 51: 1805-1812.

Arnold, W.J. and Kelley, W.N. 1973. Dietary-induced variation of hypoxanthine-guanine phosphoribosyltransferase activity in patients with the Lesch-Nyhan syndrome. J. Clin. Invest. 52: 970-973.

Bakay, B. and Nyhan, W.L. 1972. Activation of variants of hypoxanthine-guanine phosphoribosyltransferase by normal enzyme. Proc. Nat. Acad. Sci. (U.S.A.). 69: 2523-2527.

Mizuno, T., Segawa, M., Kurumada, T., Maruyama, H. and Onisawa, J. 1970. Clinical and therapeutic aspects of the Lesch-Nyhan syndrome in Japanese children. Neuropaediatrie. 2: 38-52.

Rubin, C.S., Dancis, J., Yip, L.C., Nowinski, R.C. and Balis, M.E. 1971. Purification of IMP: Pyrophosphate phosphoribosyltransferases, catalytically incompetent enzymes in Lesch-Nyhan disease. Proc. Nat. Acad. Sci. (U.S.A.) 68: 1461-1464.

Schulman, J.D., Greene, M.L., Fujimoto, W.Y. and Seegmiller, J.E. 1971. Adenine therapy for Lesch-Nyhan syndrome. Pediat. Res. 5: 77-82.

Seegmiller, J.E., Rosenbloom, F.M. and Kelley, W.N. 1967. Enzyme defect associated with a sex-linked human neurological disorder and excessive purine synthesis. Science 155: 1682-1684.

Sorensen, L.B. 1970. Mechanism of excessive purine biosynthesis in hypoxanthine-guanine phosphoribosyltransferase deficiency. J. Clin. Invest. 49: 968-978.

PROPERTIES OF ERYTHROCYTE PURINE PHOSPHORIBOSYLTRANS-
FERASES IN PARTIAL HYPOXANTHINE-GUANINE PHOSPHORIBOSYL-
TRANSFERASE DEFICIENCY

O. Sperling, P. Boer and A. de Vries

Rogoff-Wellcome Medical Research Institute and
the Metabolic Unit of Department of Medicine D,
Tel-Aviv University Medical School, Beilinson
Hospital, Petah Tikva, Israel

Partial deficiency of HGPRT, a salvage enzyme of
purine metabolism, has been demonstrated to be the
primary abnormality causing purine overproduction in a
small proportion of patients with gout (1-4). The
quantitative deviation in the activity of this enzyme
has been shown by Kelley et al. to be associated with
decreased stability to thermal inactivation (2). These
authors suggest that in the affected subjects HGPRT is
structurally altered. Furthermore, in some of these
patients erythrocyte adenine phosphoribosyltransferase
(APRT) activity was found to be increased and relatively
thermostable (2).

In the present study on a mutant HGPRT in two gouty
relatives with partial HGPRT deficiency (4), the enzyme
was found to exhibit increased sensivitivy to inhibition
by various purine nucleosides and nucleotides. On the
other hand, the APRT in these patients' erythrocytes,
except for increased activity and thermostability, did
not display any additional abnormality.

HGPRT and APRT were assayed in dialyzed hemolysates
by a radiochemical method in which labelled hypoxanthine,
guanine and adenine are converted by reaction with
phosphoribosylpyrophosphate (PRPP) to their respective
nucleotides, which then are separated by thin-layer
chromatography on microcrystalline cellulose (4).

A major difference between the mutant HGPRT and the normal enzyme could be demonstrated by exposing them to inhibition by various purine bases, nucleosides and nucleotides. The mutant enzyme was inhibited by all those purine compounds which also inhibited the normal enzyme, but several of them, namely guanosine 5'-monophosphate, guanosine 5'-diphosphate, guanosine 5'-triphosphate and inosine, inhibited the mutant enzyme to much a greater extent, particularly with guanine as substrate (Table 1). In addition, the mutant enzyme was also inhibited by guanosine which, at the concentrations used, was not inhibitory to the normal enzyme. Hypoxanthine and inosine-5-monophosphate inhibited both enzymes to the same extent, while adenine and its derivatives did not cause significant inhibition.

Slight differences were also found for other properties of the mutant HGPRT, specifically the Km for substrates, pH profile and electrophoretic mobility on starch gel. In agreement with the findings of Kelley et al. (2), the mutant enzyme exhibited increased sensitivity to heat inactivation, but similarly to the normal enzyme (5), it could be stabilized against heat by PRPP.

Regarding the APRT of these patients with partial HGPRT deficiency, all properties examined – Km for substrates, electrophoretic mobility on starch gel, pH profile and sensitivity to product and feedback inhibition, were found normal. In agreement to the findings of other investigators (2), the APRT exhibited increased activity and relative thermostability and PRPP stabilized the enzyme against thermal inactivation.

These results are interpreted to indicate a genetic structural alteration of HGPRT in this gouty family. On the other hand, regarding the APRT no evidence for a genetic alteration was obtained. The increased sensitivity of the mutant HGPRT to inhibition by various purine compounds furnishes new evidence for this enzyme being a structural mutant. The abnormal properties exhibited by the APRT, however, might be explained by the elevated PRPP content of the patient's erythrocytes.

TABLE 1.   INHIBITION OF NORMAL AND MUTANT HGPRT
BY PURINE BASES, NUCLEOSIDES AND NUCLEOTIDES

Inhibitor	Concn. (mM)	Percent residual activity of HGPRT			
		Hypoxanthine*		Guanine*	
		Normal	Mutant	Normal	Mutant
–	–	100	100	100	100
Guanine	0.023	94	71	–	–
	0.038	103	71	–	–
Guanosine	0.1	104	86	97	7.5
	0.5	114	68	99	0
GMP	0.1	62	36	43	7.5
	0.5	24	4.9	15	0
GDP	0.1	90	68	73	34
	0.5	64	35	38	5.5
GTP	0.1	89	86	83	70
	0.5	82	63	60	52

* substrate

References

1. Kelley W.N., Rosenbloom F.M., Henderson J.F. and Seegmiller J.E.   Proc. nat. Acad. Sci. (Wash.) 57:1735, 1967.

2. Kelley W.N., Greene M.I., Rosenbloom F.M., Henderson J.F. and Seegmiller J.E.   Ann. intern. Med. 70:155, 1969.

3. Kogut M.D., Donnell G.N., Nyhan W.I. and Sweetman I. Amer. J. Med. 48:148, 1970.

4. Sperling O., Frank M., Ophir R., Liberman U.A., Adam A. and De Vries A.   Europ. J. clin. biol. Res. 15:942, 1970.

5. Greene M.L., Bayles J.R. and Seegmiller J.E. Science 167:887, 1970.

RESISTANCE OF ERYTHROCYTE ADENINE PHOSPHORIBOSYL-
TRANSFERASE IN THE LESCH-NYHAN SYNDROME TO DESTABILI-
ZATION TO HEAT BY HYPOXANTHINE

P.Bashkin, O.Sperling, R.Schmidt and A.Szeinberg

Department of Chemical Pathology,Tel-Aviv
University Medical School,Tel-Hashomer, Rogoff-
Wellcome Medical Research Institute, Beilinson
Hospital,Petah-Tikva,and the Rehabilitation
Center for Children, Asaf-Harofeh Government
Hospital, Zrifin, Israel.

The Lesch-Nyhan Syndrome (LNS) is a rare x-linked
neurological disease of children characterized by
choreoathetosis, spasticity, mental retardation and
compulsive self mutilation accompanied by excessive
purine production and hyperuricemia (1). The
virtually complete deficiency of activity of a purine
salvage enzyme, hypoxanthine-guanine phosphoribosyltransferase (HGPRT) (EC 2.4.2.8.) (2), due to
structural gene mutation (3,4) has been shown to be
the basic abnormality in this disease. In erythrocytes
of LNS patients, HGPRT deficiency has been found to
be associated with increased activity and relative
thermal stability of adenine phosphoribosyltransferase
(APRT) (EC 2.4.2.7.) (5,6) an autosomally determined
enzyme (7).

A study of the properties of APRT in the
hemolysates of two children affected with LNS revealed
an additional difference between the APRT in the LNS
hemolysates and that in normal hemolysates, the LNS
enzyme being resistant to destabilization by hypoxanthine to thermal inactivation. Other properties
of the enzyme including Km for substrates, optimal
magnesium concentration, electrophoretic mobility, pH
profile and sensitivity to inhibition - were normal.

The patients were two Israeli born, 5 year old
Jewish boys of Ashkenazi (T.N.) and North African
(H.B.) origin affected with LNS (8). Dialyzed hemolysates were prepared and assayed for HGPRT and APRT
activity by measuring the incorporation of C^{14} hypoxanthine or adenine into inosinic acid or adenylic
acid respectively, as described priviously (9). The
sensitivity of APRT to thermal inactivation was
studied by measuring the residual activity of the
enzyme after heating of 20 µl dialyzed hemolysate
containing 0.11 - 0.17 mg protein at 56°C for 4 min.
The stabilization or destabilization effect of various
compounds on the thermal stability of the enzyme was
studied in the same system but in presence of 50 µM
concentration of the studied effector substances.

In the dialyzed hemolysates of the LNS children,
HGPRT activity could not be detected (< 0.05% of
normal activity) and the activity of APRT was 224% and
213% of normal respectively. In addition APRT in the
dialyzed LNS hemolysates exhibited relative stability
to thermal inactivation. Following 4 min. at 56°C, the
residual activity of the enzyme in hemolysates from
patients H.B. and T.N. was 71% and 81% respectively
(averages of 3 and 5 experiments) in comparison to
the range of 13.6 - 68.4% ($\bar{m}$ + 1SD:44.2 + 11.1%)
obtained for hemolysates of 59 control subjects.

Addition of 5-phosphoribosyl-1-pyrophosphate
(PRPP) to normal and LNS dialyzed hemolysates resulted
in stabilization of APRT to heat inactivation. Under
these conditions APRT in both normal and LNS hemolysates attained the same degree of stability
(approximately 95% residual activity).

A significant difference between the APRT in the
LNS and normal hemolysates could be demonstrated by
exposing them to destabilization against thermal
inactivation. Hypoxanthine caused a significant
decrease in the thermal stability of the enzyme in
normal hemolysates but affected only slightly the LNS
enzyme. In the presence of adenine APRT of both
normal and LNS hemolysates was equally destabilized
to thermal inactivation (Table 1). Similar findings
were recently reported also by Rubin and Balis (10).

Table 1: Stabilization and destabilization of

APRT against thermal inactivation

Compound added (50μM)	Residual enzyme activity (% of initial)		
	normal	T.N.	H.B.
-	44.2 + 11.1*	81	71
PRPP	98.8	92.9	91.8
Hypoxanthine	19.5	74.3	71.7
Adenine	13.7	9.1	11.5

* Average of 59 subjects (m + 1SD)

The mechanism by which PRPP stabilizes APRT against heat inactivation is not yet clarified, but it would be reasonable to assume that PRPP is inducing a conformational change in the APRT molecule resulting in thermal stability, and that the increased heat stability of the enzyme in the LNS hemolysate is due to the increased level of PRPP in such cells. Indeed, both increased specific activity and thermal stability of the APRT in LNS hemolysates have been attributed to stabilization of the enzyme by the elevated concentration of PRPP accumulation in the HGPRT deficient erythrocytes (11,12). In view of the fact however, that the increased stability of APRT is observed in dialyzed hemolysates, the postulated conformational change would have to be irreversible, unless the stabilizing PRPP is bound to the enzyme.

Sensitization of APRT to heat by purine bases might be due to a direct effect on the enzyme molecule, inverse to that induced by PRPP. In this case, the resistance of APRT in the LNS hemolysates to such a sensitization by hypoxanthine would be an additional property of the PRPP-induced conformationally-altered APRT molecule. On the other hand if the stability of APRT to heat is dependent on the presence of enzyme-bound PRPP, the depletion of the latter would also cause destabilization. This could occur if the APRT bound PRPP is available for enzymatic reactions with hypoxanthine and adenine, catalyzed by HGPRT and APRT respectively. In this case the resistance of APRT in the LNS hemolysates to destabilization by hypoxanthine would be merely the result of HGPRT deficiency in such cells.

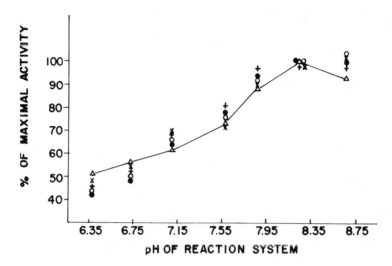

Fig. 1   pH profile of APRT activity in hemolysates
          △ - normal subjects; ○ and ● LNS patients;
          x and + mothers of LNS patients.

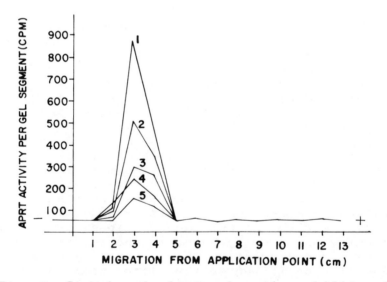

Fig. 2   Starch gel electrophoretic mobility of
          APRT.   1 and 2 - LNS patients; 3 and 4
          - mothers of LNS patients; 5 - normal
          subjects.

Table 2: Kinetic properties of APRT

Subjects	Km for adenine (μM)	Km for PRPP (μM)	$\left[Mg^{++}\right]_{0.5}$* (mM)
Normal (6 subjects)	1.14 - 2.3	3.7 - 5.7	0.36 - 0.38
LNS Subjects			
H.B.	1.34	3.2	0.37
T.N.	1.42	3.9	0.37

* Mg^{++} concentration at one half maximal activity of the enzyme.

Both of the above postulated mechanisms for the explanation of abnormalities exhibited by the APRT in the dialyzed LNS hemolysates are consistent with the assumption that they are not genetic, but secondary to HGPRT deficiency and the resulting accumulation of PRPP. The finding that the APRT in the LNS hemolysate has normal **Km**'s for substrate, normal activation by magnesium (Table 2), normal pH profile (Fig. 1) and normal electrophoretic mobility (Fig. 2), is also consistent with this assumption.

REFERENCES

1. Lesch M. and Nyhan WL.
    Am.J.Med. 36 : 561, 1964.

2. Seegmiller JE. Rosenbloom FM and Kelley WN.
    Science 155 : 682, 1967.

3. Fujimoto WY and Seegmiller JE.
    Proc.Nat.Acad.Sci. 65 : 577, 1970.

4. Kelley WN and Meade JC.
    J.Biol.Chem. 246 : 2953, 1971.

5. Kelley WN. Greene ML. Rosenbloom FM. Henderson JF
    and Seegmiller JE.
    Ann.Intern.Med. 70 : 155, 1969.

6. Kelley WN.
    Fed.Proc. 27 : 1047, 1968.

7.   Henderson JF. Kelley WN. Rosenbloom FM and
     Seegmiller JE.
     The Amer.J.Human Genetics. 21 : 61, 1969.

8.   Schmidt R. Mundel G. and Sperling O.
     Harefuah 82 : 410, 1972.

9.   Sperling O. Frank M. Ophir R. Liberman UA. Adam A.
     and De-Vries A.
     Europ.J.Clin.Biol.Res. 15 : 942, 1970.

10. Rubin CS. and Balis ME.
     Biochem.Biophys.Acta 279 : 163, 1972.

11. Rubin CS. Balis ME. Piomelli S. Berman PH and
     Dancis J.
     J.Lab.Clin.Med. 74 : 732, 1969.

12. Greene ML. Bayles JR and Seegmiller JE.
     Science 167 : 887, 1970.

Purine Metabolism in HGPRT Deficient Cells

PURINE METABOLISM IN INTACT ERYTHROCYTES FROM CONTROLS AND HG-PRT

DEFICIENT INDIVIDUALS

C.H.M.M. de Bruyn and T.L. Oei
Department of Human Genetics, Faculty of Medicine
University of Nijmegen, The Netherlands

The present paper reports on studies on the purine metabolism
in normal and HG-PRT deficient human erythrocytes.

Intact normal and HG-PRT deficient human erythrocytes were
incubated in the presence of radioactively labeled purine bases,
nucleosides and nucleotides, respectively. The uptake, interconversion and release of labeled compounds were studied by analysis
of both medium and TCA supernatant of the red cells (cell content)
by means of paper chromatography and high-voltage electrophoresis,
followed by liquid scintillation counting.

Additionally, the enzymes involved in the purine metabolism
in human erythrocytes were assayed in the 30.000 g supernatants
of haemolyzed red cells, as well as in the carefully washed 30.000
g pellet (cell membranes) using radioactively labeled substrates.

From the large amounts of data obtained in this way only a
small part is presented.

Interesting was the finding of HG-PRT and A-PRT activities
in red cell membranes. This has not been reported before (Slide 1,
Table I).

Preliminary studies including pH optimum, influence of metal
ions and substrate affinities revealed no essential differences
between the supernatant enzymes and the membrane-bound enzymes.
However, the heat stability of membrane-bound A-PRT is much higher
than supernatant A-PRT.

It is known that A-PRT is stabilized by PRPP. Therefore, the
observation that membrane-bound A-PRT and HG-PRT show considerable
activity without supply of exogenous PRPP to a thoroughly washed
pellet, is suggestive of the presence of PRPP in the red cell membrane, associated with A-PRT and HG-PRT.

In erythrocytes obtained from Lesch-Nyhan patients HG-PRT activity was also lacking in the cell membrane. This observation and

223

	Purine phosphoribosyltransferase activity nmol/mg protein/hour		
	H-PRT	G-PRT	A-PRT
Erythrocyte 30.000 g sup.	70.6	123.3	17.9
Erythrocyte 30.000 g pellet	66.9	97.8	89.8

Table I: Purine phosphoribosyltransferase activities in 30.000 g

supernatants and washed pellets obtained from erythro-

cyte lysates.

the similarity of enzyme properties seems to indicate that the
membrane-bound and supernatant enzymes are the same.

The membrane-bound HG-PRT and A-PRT activities could also be
demonstrated in the intact erythrocytes. Normal red cells were incubated with ^{14}C-hypoxanthine, ^{14}C-guanine and ^{14}C-Adenine, respectively, in the presence or absence of PRPP in the incubation
medium.

Slide 2 (Table II) shows the results: PRPP not only accele-
rates the IMP, GMP and AMP formation inside the red cells, but it
also induces the appearance of these nucleotides outside the red
cells. This shows that membrane-bound HG-PRT and A-PRT are acces-
sible from the outside of the cell. It also indicates that these
enzymes have some role in the transport of purines across the cell
membrane.

Results of incubation experiments with intact normal and HG-
PRT deficient cells support the theory (ref.1,2), that HG-PRT deficiency causes decreased levels of adenine nucleotides in the red
blood cell.

Slide 3 (Fig.1) shows that adenosine from the medium is rapidly converted to inosine,probably on the cell membrane, since no adenosine is detectable inside the cell. Inosine is rapidly converted
to IMP in the normal cell,but in mutant cells it is converted to
hypoxanthine. Effective AMP formation from adenosine in mutant
cells is lower than in normal cells, indicating a shift towards
hypoxanthine formation.

	14C-Hypoxanthine added to medium (0,01 mM)		14C-Guanine added to medium (0,01 mM)		14C-Adenine added to medium (0,01 mM)	
	without PRPP	+5mM PRPP in medium	without PRPP	+5mM PRPP in medium	without PRPP	+5mM PRPP in medium
medium	Hx:8853 cpm IMP: ——	Hx: —— IMP:8726 cpm	Gu:7790 cpm GMP: ——	Gu: —— GMP:7655 cpm	Ad:9114 cpm AMP: ——	Ad:7456 cpm AMP:1304 cpm
cells	Hx:3654 cpm IMP:4771 cpm	Hx:1383 cpm IMP:7597 cpm	Gu:3270 cpm GMP:3288 cpm	Gu: —— GMP:6547 cpm	Ad:5269 cpm AMP:2601 cpm	Ad:3775 cpm AMP:4055 cpm
total cpm	17.278	17.706	14.348	14.202	16.984	16.590

Table II: Metabolisme of purine bases by intact human erythrocytes in the absence and presence of exogenous PRPP.

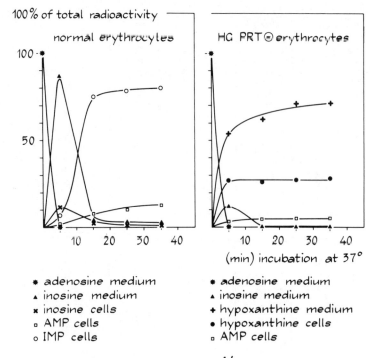

* adenosine medium   * adenosine medium
▲ inosine medium   ▲ inosine medium
✕ inosine cells   + hypoxanthine medium
□ AMP cells   ● hypoxanthine cells
○ IMP cells   □ AMP cells

Fig.1: Uptake and interconversions of ^{14}C-adenosine by intact normal and HG-PRT erythrocytes.

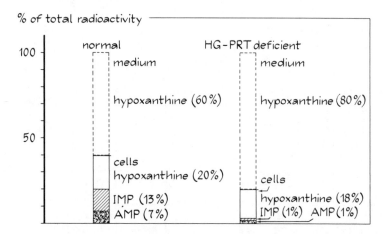

Fig.2: Recovery of the label administered as ^{14}C-adenine from incubations with intact normal and HG-PRT deficient erythrocytes.

This model is also supported by the results shown in slide 4 (Fig.2). Normal and HG-PRT deficient erythrocytes were washed and incubated with ^{14}C-adenine in autologous serum. In normal cells, along with IMP and AMP, the major endproduct was hypoxanthine. In normal cells IMP and AMP were formed in considerable amounts, but in mutant cells only traces of IMP and AMP were recovered. This illustrates the importance of the salvage pathway involving HG-PRT.

It has been suggested (ref.3) that increased IMP dehydrogenase activity in HG-PRT deficient cells could be responsible for a shift from IMP towards GMP. This might explain why in HG-PRT deficient cells normal concentrations of guanine nucleotides are encountered (ref.4). Under our conditions,however,no XMP or GMP were observed, even after incubations up to 24 hours. These findings cast some doubt on the physiological significance of IMP dehydrogenase in the intact red cell.

Acknowledgements

This study was supported by a grant of FUNGO, Foundation for Medical Scientific Research in the Netherlands.
The authors thank Mr. C. van Bennekom and A.Janssen for skillfull technical assistance.

References

1. M.E. Balis (1968) : Aspects of purine metabolism, Fed.Proc. 27, 1067.
2. L.B. Sorensen (1970) : Mechanism of excessive purine biosynthesis in hypoxanthine-guanine phosphoribosyltransferase deficiency, J. Clin. Invest. 49, 968.
3. D.M. Pehlke, J.A. McDonald, E.W. Holmes and W.N. Kelley (1972): Inosinic acid dehydrogenase activity in the Lesch-Nyhan syndrome, J. Clin. Invest. 51, 1398.
4. E.J.P. Lommen, G.D. Vogels, S.P.M. van der Zee, J.M.F. Trijbels and E.D.A.M. Schretlen (1971): Concentration of purine nucleotides in erythrocytes of patients with the Lesch-Nyhan syndrome before and during oral administration of adenine, Acta Paediat. Scand. 60, 642.

INCORPORATION OF ^{3}H-HYPOXANTHINE IN PHA[x] STIMULATED HG-PRT

DEFICIENT LYMPHOCYTES

C.H.M.M. de Bruyn and T.L. Oei
Department of Human Genetics, Faculty of Medicine
University of Nijmegen, The Netherlands

Autoradiography with labeled substrates on cultured fibroblasts has been used to visualize HG-PRT deficiency: cells of deficient individuals lack the ability to incorporate hypoxanthine or guanine into nucleic acids, because these purine bases can not be converted to their corresponding mononucleotides (1,2). The alternative pathway to form IMP or GMP via inosine or guanosine is not likely, for, although nucleoside phosphorylase is present, there is no definite evidence for the existence of inosine- or guanosine kinase in human cells (3,4).

In this paper autoradiographic experiments on lymphocytes obtained from several unrelated HG-PRT deficient individuals and from a control are reported. Also some preliminary experiments to quantify the autoradiographic data are described.

From 20 ml heparinized venous blood lymphocytes were isolated and one part of these was suspended in TC 199 medium, supplemented with calf serum, chicken embryo extract, antibiotics, and cultured for 72 hours in the presence of phytohaemagglutinin (PHA). The other part was used for experiments immediately. Before incubation, the cells were washed twice with isotonic Na,K-phosphate buffer pH 7.4 and resuspended in incubation medium, containing hypoxanthine-8-^{3}H (10 /uCi/ml). After incubation for three hours at 37º, cells were spun down and washed three times. After hypotonic treatment and fixation, the cells were brought on slides, spread by flame drying, and stained with orceïne. The slides were coated with Kodak Nuclear Track emulsion and developed after exposure for 7 days. Details of the procedure are reported elsewhere (5).

[x] PHA = phytohaemagglutinin.

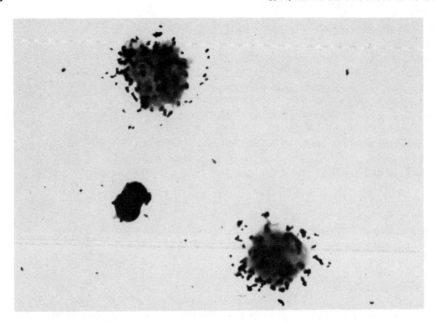

Fig.1: Unstimulated normal lymphocytes after incubation with ^{3}H-hypoxanthine.

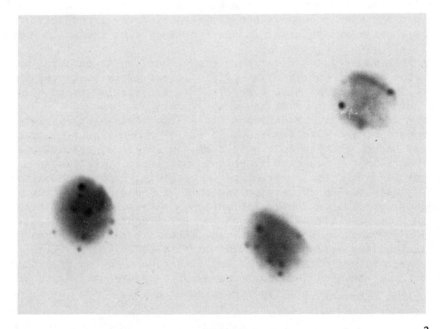

Fig.2: Unstimulated HG-PRT lymphocytes after incubation with ^{3}H-hypoxanthine.

Lymphocytes of a control show heavy labeling (Slide 1, Fig.1), whereas very little label is observed in the Lesch-Nyhan lymphocytes, as could be expected (Slide 2, Fig.2). In normal stimulated cells also a heavy labeling was seen (Slide 3, Fig.3). Surprisingly, however, the stimulated Lesch-Nyhan lymphoblasts were heavily labeled too.

Slides 4 and 5 (Figs. 4 and 5) show results with cells of unrelated HG-PRT deficient individuals. The labeled hypoxanthine, a known precursor of nucleic acids, is incorporated into chromosomal material: this can be concluded from slide 6 (Fig.6), were a metaphase with labeled chromosomes is shown in a PHA stimulated white blood cell from a HG-PRT deficient individual.

To quantify the incorporation of purine bases in TCA principitable macro molecules, PHA stimulated and unstimulated HG-PRT negative cells were incubated with ^{14}C-hypoxanthine and ^{14}C-guanine. The preliminary results are summarized in table I.

In unstimulated cells some labeling of the TCA pellet was found; no increase of radioactivity occurred during prolonged incubation. In the TCA pellets of PHA stimulated cells, however, increasing amounts of incorporated guanine and hypoxanthine were observed after longer periods of incubation (table I). These data confirm the autoradiographic results.

Table II shows the percentage of cells labeled after several periods of incubation with PHA. After 9 hours, significant

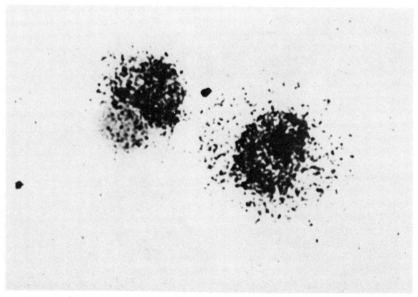

Fig.3: PHA stimulated normal lymphocytes after incubation with
3H-hypoxanthine.

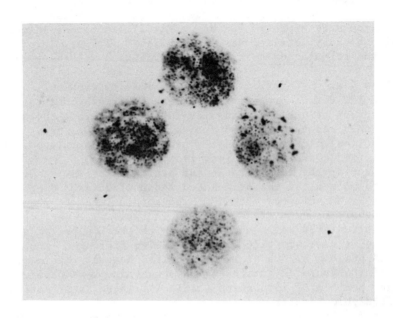

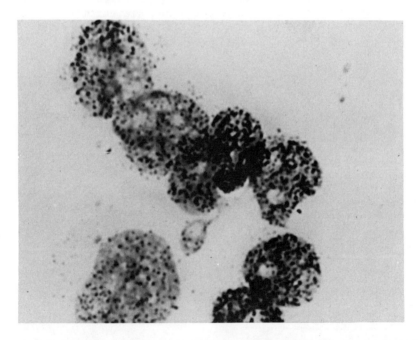

Figs. 4 and 5: PHA stimulated HG-PRT lymphocytes from unrelated
Lesch-Nyhan patients after incubation with ^{3}H-hypoxanthine.

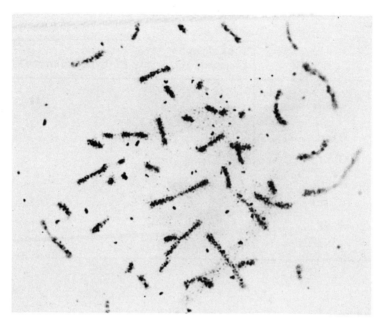

Fig.6: Labeled metaphase chromosomes from PHA stimulated HG-PRT⁻ lymphocytes.

Table I: Radioactivity in TCA precipitates of PHA stimulated and unstimulated HG-PRT⁻ lymphocytes after incubation with ^{14}C-guanine and ^{14}C-hypoxanthine.

	cmp in TCA pellet/mg protein after incubation with:	
	^{14}C-guanine (0,065 mM)	^{14}C-hypoxanthine (0,065 mM)
unstimulated lymphocytes		
t = 2 hr	3495	1171
t = 4 hr	2860	1011
PHA stimulated lymphocytes		
t = 0,5 hr	3570	1573
t = 1 hr	5027	2068
t = 3 hr	9752	2987

Table II: Percentage of labeled cells after various periods
following PHA administration (incubations with ^{3}H-hypoxanthine)

	% of lymphocytes labeled (incubation with ^{3}H-hypoxanthine)
time after addition of PHA	
0 hr	—
3 hr	—
9 hr	8%
18 hr	9%
24 hr	14%
72 hr	97%

numbers (8%) of cells contained TCA precipitable label. After 72
hours almost all cells (97%) were labeled. It is known that during
the first few hours after its addition, PHA stimulates cytoplasmic
RNA synthesis. After 6 hours an increase in ribosomal RNA synthesis
can be observed (ref.6), followed by high rates of protein synthesis
(ref.7). The synthesis of DNA begins at about 24 hr. To relate the
results in table II with the processes during PHA stimulation, more
detailed studies are needed.

     Possible explanations for the capacity to incorporate guanine
and hypoxanthine after PHA stimulation are:
a: HG-PRT is induced or activated;
b: inosine kinase is induced or activated;
c: some unknown pathway of purine metabolism is opened up in PHA
   stimulated HG-PRT deficient lymphocytes.
Experiments to elucidate the mechanism of the PHA induced incorporation of purine bases are in progress.

Acknowledgements

     This study was supported by FUNGO(Foundation for Medical
Scientific Research in the Netherlands).The skillfull technical
assistance of Mrs. I. Janssen-Hoos, Mr. A.Janssen and Mr. C. van
Bennekom is gratefully acknowledged.

References

1. B.R. Migeon, W.M. Der Kaloustian, W.M. Nyhan, W.Y. Young and
   B. Childs (1968).
   X-linked hypoxanthine-guanine phosphoribosyltransferase defi-
   ciency: heterozygote has two clonal populations.
   Science 160, 425.
2. J. Dancis, R.P. Cox, P.H. Berman, V. Jansen and M.E. Balis (1969).
   Cell population density and phenotypic expression of tissue
   culture fibroblasts from heterozygotes of Lesch-Nyhan's disease.
   Biochem. Genet. 3, 609.
3. M.R. Payne, J. Dancis, P.H. Berman and M.E. Balis (1970).
   Inosine kinase in leucocytes of Lesch-Nyhan patients.
   Exptl. Cell Res. 59, 489.
4. T. Friedmann, J.E. Seegmiller and J.H. Subak-Sharpe (1969).
   Evidence against the existence of guanosine and inosine kinases
   in human fibroblasts in tissue culture.
   Exptl. Cell Res. 56, 425.
5. C.H.M.M. de Bruyn and T.L. Oei (1973).
   Lesch-Nyhan syndrome: incorporation of hypoxanthine in stimu-
   lated lymphocytes.
   Exptl. Cell Res. in press.
6. J.E. Kay (1968).
   Early effects of phytohaemagglutinin on lymphocyte RNA synthesis.
   Eur. J. Biochem. 4, 225.
7. G.C. Mueller and M.L. Mahien (1966).
   Induction of RNA synthesis in human leucocytes by phytohaemag-
   glutinin.
   Biochim. Biophys. Acta 114, 100.
8. C.H.M.M. de Bruyn and T.L. Oei (1973).
   Purine metabolism in intact erythrocytes from control and HG-PRT
   deficient individuals.
   These Proceedings.

STUDIES ON METABOLIC COOPERATION USING DIFFERENT TYPES OF NORMAL

AND HYPOXANTHINE-GUANINE PHOSPHORIBOSYLTRANSFERASE (HG-PRT)

DEFICIENT CELLS

T.L. Oei and C.H.M.M. de Bruyn
Department of Human Genetics, Faculty of Medicine
University of Nijmegen, The Netherlands

The experiments reported in this paper were designed to test
one of several possible mechanisms that may underly the phenomenon
of metabolic cooperation between normal and HG-PRT deficient cells,
namely transfer of some intermediate of the purine metabolism from
cell to cell.

Since Subak-Sharpe et al. (1,2) described the results of autoradiographic studies on mixed cultures of HG-PRT positive and negative cell lines, many workers have studied metabolic cooperation in
a similar system, i.e. in mixtures of cultured normal and enzyme
deficient fibroblasts. Drawback of such a system is the difficulty
to separate normal and enzyme deficient cells afterwards.

Our aim was to analyse the process of metabolic cooperation separately in the HG-PRT positive and negative cells before and after
interaction of the cells. Therefore, the feasibility of using different types of normal and enzyme deficient cells in experiments
on metabolic cooperation was investigated.

First, freshly isolated lymphocytes obtained from a normal person were preincubated in ^{3}H-hypoxanthine containing medium, and then
washed out several times. The last washing fluid was shown to contain no measurable radioactivity by liquid scintillation counting.
The preincubated normal lymphocytes were then mixed with an approximately equal amount of untreated lymphocytes obtained from a Lesch-
Nyhan patient. The cells were spun down to assure cellular contact.
The cells were then fixed, treated with cold TCA, stained with or-
ceine and subjected to autoradiography.

Slide 1 (Fig.1) shows the results of this experiment. All cells
contain ^{3}H-labeled material. Evidently some label has been transferred from normal to enzyme deficient lymphocytes. There are two
classes of cells: some are more, some less heavily labeled. This
reflects the difference between the preincubated normal lymphocytes

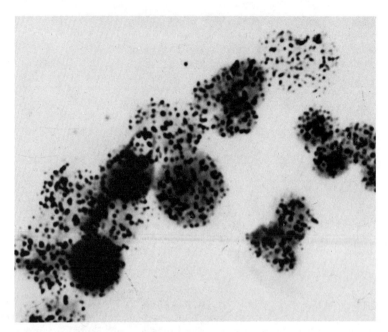

Fig.1: TCA precipitable label in both normal lymphocytes preincubated with ^{3}H-hypoxanthine, and HG-PRT$^-$ lymphocytes not preincubated with ^{3}H-hypoxanthine.

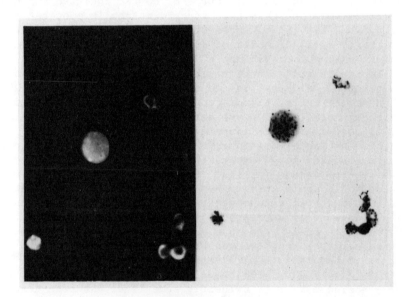

Fig.2: F-body and TCA precipitable label in a lymphocyte from a
Lesch-Nyhan patient, after interaction with ^{3}H-hypoxanthine preincubated lymphocytes from a normal female donor.

and the enzyme deficient cells that have not been preincubated with
^{3}H-hypoxanthine.

The experiment was repeated with preincubated lymphocytes from
a normal female donor. Before autoradiography was performed, the
HG-PRT deficient cells were identified by staining with quinacrine
mustard.

Slide 2 (Fig.2) shows a lymphocyte containing a fluorescent F-
body, i.e. a Y chromosome, and the same cell after autoradiography.
This definitely proves that the HG-PRT deficient lymphocytes have
somehow received and incorporated the label.

Next, erythrocytes from a normal donor were preincubated in
the presence of ^{3}H-hypoxanthine and then mixed with untreated HG-PRT
negative lymphocytes. The advantage of using erythrocytes is that
the purine metabolism is much simpler in these cells: there is no
DNA and RNA synthesis, or de novo synthesis of purines. Additionally,
the erythrocytes can be readily removed from mixtures with other
types of cells by hypotonic treatment.

Slide 3 (Fig.3) shows the incorporated label in the HG-PRT negative lymphocytes, after cellular contact with preincubated normal
erythrocytes and removal of the red cells by hypotonic treatment,
followed by autoradiography. It demonstrates transfer of labeled
material to the lymphocytes.

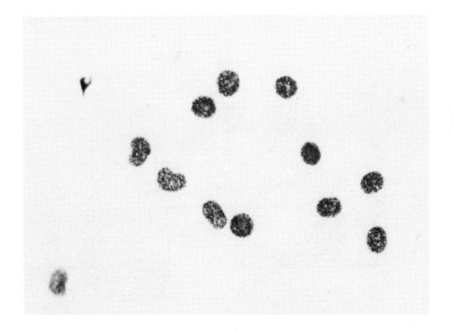

Fig.3: TCA precipitable label in HG-PRT$^-$ lymphocytes after interaction with ^{3}H-hypoxanthine preincubated normal erythrocytes.

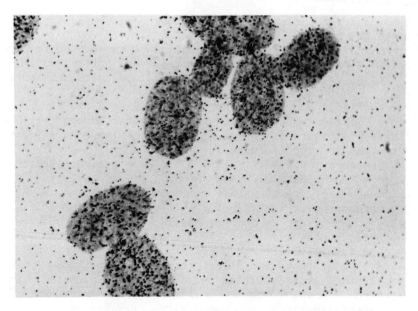

Fig.4: TCA precipitable label in HG–PRT⁻ fibroblasts after incubation with ^{3}H-hypoxanthine preincubated normal erythrocytes.

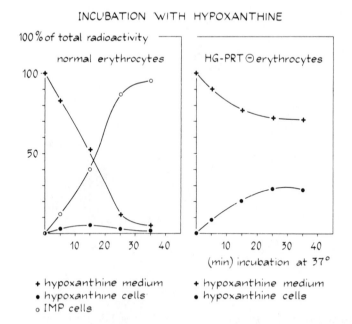

Fig.5: Uptake and interconversion of ^{3}H-hypoxanthine by normal
and HG–PRT⁻ erythrocytes.

Slide 4 (Fig.4):Here HG-PRT negative human skin fibroblasts
were grown on coverslips. Normal erythrocytes preincubated with
^{3}H-hypoxanthine were added to the culture medium and were allowed
to settle down on top of the fibroblasts. The erythrocytes were
then poured off, and the few remaining red cells lysed by hypotonic
shock. Autoradiography reveals incorporation of radioactive label
in the HG-PRT negative fibroblasts.

Studies on the purine metabolism in intact normal and HG-PRT
deficient human red cells (3) showed that the erythrocytes, which
were used in the experiments on metabolic cooperation, contained
the radioactive label almost exclusively in the form of IMP:Slide 5
(Fig.5). The donor erythrocytes in the transfer experiments were
preincubated during 40 min. After that treatment, besides trace
amounts of hypoxanthine, only IMP was recovered from the red cell
content. In these transfer experiments the preincubated normal erythrocytes were small packets containing ^{3}H-IMP.

It is generally known, however, that IMP does not penetrate
the cell membrane and this was confirmed experimentally: HG-PRT deficient fibroblasts following incubation in medium containing ^{3}H-
IMP do not show any incorporation of label after autoradiography.
Surprisingly, addition of cell membrane fractions from both normal
and HG-PRT deficient fibroblasts or erythrocytes to the ^{3}H-IMP containing medium induces incorporation of label into the HG-PRT deficient fibroblasts: Slide 6 (Fig.6).

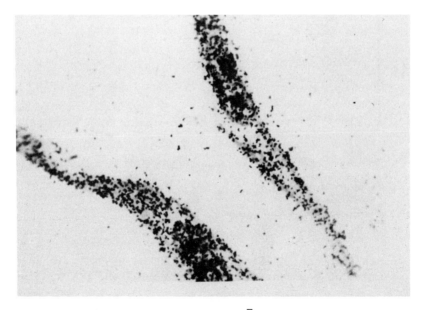

Fig.6: TCA precipitable label in HG-PRT⁻ fibroblasts incubated
with ^{3}H-IMP in the presence of cell membranes.

Finally, HG-PRT negative erythrocytes were incubated with ^{3}H-IMP in the presence and absence of cell membrane fractions of normal or enzyme deficient fibroblasts, respectively. In the presence of cell membrane fractions ^{3}H-IMP is recovered from the HG-PRT negative red cell content; in the absence of cell membrane fractions this is not the case.

The experiment was repeated with ^{3}H-GMP: Slide 7 (Fig.7).It is interesting to note that in this system, i.e. in the presence of membrane fractions, GMP appears about 3 x faster in the enzyme deficient red cell than IMP.

Concluding and summarizing this communication:
1. Experimental evidence is presented that nucleotides (IMP and GMP) can be transferred from normal to HG-PRT negative cells.
2. The results reported indicate that membrane-membrane interaction is presumably essential in this type of metabolic cooperation.

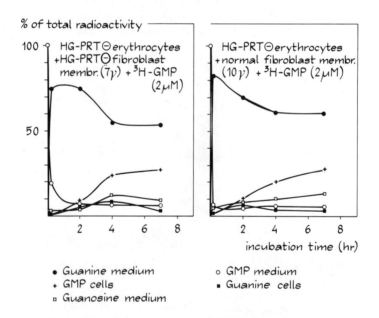

Fig.7: Uptake and interconversion of ^{3}H-GMP by HG-PRT$^-$ erythrocytes in the presence of membrane fractions of normal and HG-PRT$^-$ fibroblasts.

Acknowledgements

This investigation was in part supported by FUNGO (Foundation for Medical Scientific Research in the Netherlands).

The authors thank Mrs. I. Janssen-Hoos, Mrs. C. Raymakers-Volaart, Mr. C. van Bennekom, and Mr. A. Janssen for expert technical assistance. Prof. Dr. S.J. Geerts is gratefully acknowledged for helpfull discussions.

References

(1) H. Subak-Sharpe, R.R. Bürk and J.D. Pitts (1966):
Metabolic cooperation by cell to cell transfer between genetically different mammalian cells in tissue culture.
Heredity 21, 342-343.
(2) H. Subak-Sharpe, R.R. Bürk and J.D. Pitts (1969):
Metabolic cooperation between biochemically marked mammalian cells in tissue culture.
J. Cell Sci. 4, 353-367.
(3) C.H.M.M. de Bruyn and T.L. Oei (1973):
Purine metabolism in intact erythrocytes from controls and HG-PRT deficient individuals.
These Proceedings.

REGULATION OF DE NOVO PURINE SYNTHESIS IN THE LESCH-NYHAN SYNDROME

Gabrielle H. Reem, M.D.

Department of Pharmacology, New York University Medical

Center, N.Y.

The deficiency of hypoxanthine guanine-phosphoribosyltransferase (HGPRT) activity of patients with the Lesch-Nyhan syndrome is accompanied by an accelerated rate of purine biosynthesis de novo. This elevated rate of purine biosynthesis is reflected in fibroblasts cultured from the skin of these patients and is accompanied by elevated levels of phosphoribosylpyrophosphate (PRPP).

The purpose of this study was to examine the regulation of de novo purine biosynthesis in cultured lymphocytes from Lesch-Nyhan patients and to compare it with that of cultured lymphocytes of a normal subject. Two cell lines of lymphocytes were studied: One was established from a normal 26-year-old female, the other from a 3-year-old boy with the Lesch-Nyhan syndrome. The Lesch-Nyhan lymphocytes were deficient in HGPRT activity, and their PRPP concentration was 3-6 times higher than that of the control lymphocytes.

Purine biosynthesis was studied in intact cells maintained in culture and in cell free preparations. In cell-free preparations the regulation of phosphoribosylamine (PRA) synthesis was studied, since this step is believed to be the rate-limiting step in de novo purine biosynthesis. PRA can be synthesized according to the following reactions:

1. PRPP + glutamine + H_2O
2. PRPP + ammonia ⟶ PRA
3. Ribose-5-P + ammonia + ATP

The regulation of PRA synthesis from PRPP and glutamine (reaction
1) and PRPP and ammonia (reaction 2) was examined in cell-free
preparations of human lymphocytes. (1-5)

STUDIES IN LYMPHOCYTES IN CULTURE

The rate of the early steps of de novo purine biosynthesis of
lymphocytes in culture was measured by determining the incorporation of labeled glycine into ^{14}C-α-N-formylglycinamide ribonucleotide (^{14}C-FGAR) of azaserine treated cells.

TABLE 1. RATE OF THE EARLY STEPS OF PURINE BIOSYNTHESIS IN HUMAN
LYMPHOCYTES IN CULTURE

	ADDITIONS		^{14}C-FGAR
			cpm/10^6 cells
Normal Lymphocytes	None		104
	Glutamine	0.4 mM	430
	NH$_4$Cl	10.0 mM	488
Lesch-Nyhan Lymphocytes	None		69
	Glutamine	0.4 mM	647
	NH$_4$Cl	10.0 mM	558

Since either glutamine or ammonia can serve as a substrate for PRA
synthesis, the addition of either glutamine or ammonia to the incubation medium resulted in stimulation of the rate of FGAR synthesis in intact lymphocytes. The basal rate of labeled glycine incorporation into FGAR was 104 CPM/hour/10^6 cells in normal lymphocytes in culture, and 69 CPM/hour/10^6 cells in Lesch-Nyhan cells.
Addition of glutamine (0.4 mM) to the incubation medium of normal
lymphocytes increased FGAR synthesis to 430 CPM/hour/10^6 cells,
while the addition of glutamine to Lesch-Nyhan cells raised the
rate of FGAR synthesis to 647 CPM/hour/10^6 cells. The addition of
10 mM NH$_4$Cl (equivalent to 0.15 mM ammonia at pH 7.5) stimulated
the incorporation of ^{14}C-glycine into ^{14}C-FGAR to 488 CPM/hour/10^6
cells in normal lymphocytes, and to 588 CPM/hour/10^6 cells in
Lesch-Nyhan lymphocytes.

The regulation of the early steps of de novo purine biosynthesis was examined by studying the effect of the addition of purine bases to the incubation medium (Figure 1). Adenine was a potent inhibitor of FGAR synthesis both in the Lesch-Nyhan lymphocytes and in the normal lymphocytes. In normal lymphocytes 0.5 micromolar adenine inhibited ammonia dependent FGAR synthesis by 50%; in Lesch-Nyhan lymphocytes 1.4 micromolar adenine resulted in 50% inhibition of FGAR synthesis.

Normal lymphocytes were also very sensitive to the inhibitory effect of hypoxanthine (Figure 1). The addition of 2 micromolar hypoxanthine to the incubation medium caused 50% inhibition of ammonia-dependent FGAR synthesis in normal lymphocytes. In Lesch-Nyhan lymphocytes 41 times that amount of hypoxanthine (0.8 mM) was required to attain 50% inhibition of ammonia-dependent FGAR synthesis. When glutamine replaced ammonia in the incubation medium the inhibitory effect of adenine and hypoxanthine on FGAR synthesis resembled that observed with ammonia (Figure 2). 0.6 micromolar adenine inhibited glutamine-dependent FGAR synthesis of normal lymphocytes by 50% and 1.2 micromolar adenine inhibited FGAR synthesis of Lesch-Nyhan lymphocytes by 50%. The HGPRT deficient Lesch-Nyhan cells were relatively resistant to inhibition by hypoxanthine. While 4.2 micromolar hypoxanthine inhibited FGAR synthesis of normal lymphocytes by 50%, 65 micromolar hypoxanthine was necessary to achieve the same effect in Lesch-Nyhan lymphocytes.

STUDIES IN CELL-FREE PREPARATIONS

PRA synthesis from PRPP and glutamine and from PRPP and ammonia in cell-free preparations of Lesch-Nyhan lymphocytes was compared with that of cell-free preparations of normal lymphocytes (Table 2). In preparations of normal lymphocytes 1.88 nmoles PRA/min/mg protein were synthesized from PRPP and glutamine and 7.5 nmoles PRA/min/mg was formed from PRPP and ammonia. In preparations of Lesch-Nyhan cells 2.76 nmoles PRA/min/mg was formed from PRPP and glutamine, and 12.5 nmoles PRA/min/mg from PRPP and ammonia.

Effect of Purine Bases

The effect of purine bases on PRA synthesis from PRPP and glutamine and from PRPP and ammonia was studied in cell-free preparations of Lesch-Nyhan lymphocytes and compared with that of normal lymphocytes (Table 3). 0.5 mM adenine was an effective in-

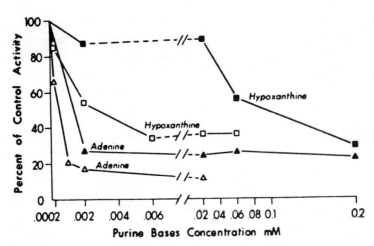

Fig.1   Effect of Purine Bases on the Rate of Ammonia Dependent
Purine Synthesis of Human Lymphocytes in Culture.
△ □, Normal human lymphocytes; ▲ , ■,Lesch-Nyhan
lymphocytes.

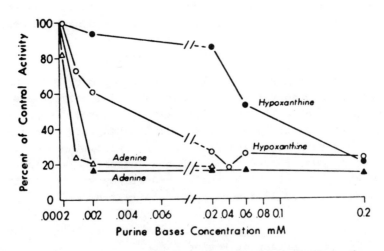

Fig.2   Effect of Purine Bases on the Rate of Glutamine
Dependent Purine Synthesis of Human Lymphocytes in Culture.
△, ○, Normal human lymphocytes;   ▲, ●,Lesch-Nyhan
lymphocytes.

TABLE 2.   PRA SYNTHESIS IN EXTRACTS OF HUMAN LYMPHOCYTES IN CULTURE

	SUBSTRATES	
	AMMONIA and PRPP	GLUTAMINE and PRPP
	nmoles PRA/min/mg protein	
Normal Lymphocytes	7.5	1.88
Lesch-Nyhan Lymphocytes	12.5	2.76

hibitor of these two reactions in preparations of normal and of
Lesch-Nyhan lymphocytes. Hypoxanthine did not inhibit PRA synthesis from either PRPP and glutamine or from PRPP and ammonia in
cell-free preparations of the HGRPT deficient Lesch-Nyhan lymphocytes while in preparations of normal lymphocytes hypoxanthine inhibited PRA synthesis effectively.

TABLE 3.   EFFECT OF PURINE BASES ON PRA SYNTHESIS IN EXTRACTS OF
LESCH-NYHAN LYMPHOCYTES AND OF NORMAL LYMPHOCYTES

PRPP CONCENTRATION	INHIBITOR	GLUTAMINE PRPP AMIDOTRANSFERASE		AMMONIA PRPP AMINOTRANSFERASE	
		L.N.	Normal	L.N.	Normal
		% inhibition			
.25 mM	Adenine	65	35	84	76
.5 mM	(0.5 mM)	48	47	69	
1.0 mM		52	79	52	58
.25 mM	Hypoxanthine	6	61	23	64
.5 mM	(0.5 mM)	4	79	6	50
1.0 mM		6	56	10	7

Effect of Purine Ribonucleotides

The study of the effect of purine ribonucleotides on PRA synthesis in cell free enzyme preparations of these two lines of lymphocytes revealed that PRA synthesis of preparations of HGPRT deficient Lesch-Nyhan cells were sensitive to inhibition by AMP, GMP and allopurinol ribonucleotide (Table 4). The degree of inhibition of PRA synthesis of preparations of normal lymphocytes by AMP, GMP and allopurinol ribonucleotide was similar to that of preparations of Lesch-Nyhan lymphocytes. There was, however, a significant difference between the effect of AMP on PRA synthesis from PRPP and glutamine compared with that on PRA synthesis from PRPP ammonia. PRA synthesis from PRPP and ammonia was significantly more sensitive to inhibition by AMP than was PRA synthesis from PRPP and glutamine.

TABLE 4.   EFFECT OF PURINE RIBONUCLEOTIDES ON PRA SYNTHESIS IN
           EXTRACTS OF LESCH-NYHAN AND NORMAL LYMPHOCYTES

PRPP CON-CENTRATION	INHIBITOR	GLUTAMINE PRPP AMIDOTRANSFERASE		AMMONIA PRPP AMINOTRANSFERASE	
		L.N.	Normal	L.N.	Normal
			% inhibition		
.25 mM	AMP	30	57	78	69
.5 mM	(0.5 mM)	28	34	65	62
1.0 mM		19	13	75	47
.25 mM	GMP	82	52	85	80
.5 mM	(0.5 mM)	55	21	65	86
1.0 mM		25	32	64	81
.25 mM	Allopurinol	64	82	88	76
.5 mM	Ribonucleotide	40	72	40	35
1.0 mM	(0.5 mM)	29	60	84	

This difference in response to inhibition by AMP was observed in both cell lines; it is more clearly illustrated on Fig. 3. AMP (0.5 mM) inhibited PRA synthesis utilizing ammonia as the source of nitrogenous substrate by 65% to 78% in preparations of Lesch-Nyhan cells and by 47% to 69% in preparations of normal lymphocytes at PRPP concentrations from 0.25 mM to 1 mM, while PRA synthesis from PRPP and glutamine was inhibited by 19% to 30% in Lesch-Nyhan preparations and 13% to 57% in control preparations (Figure 3).

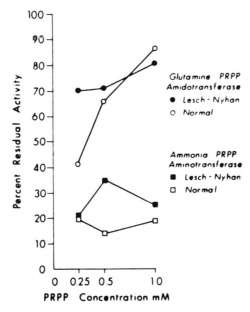

Fig. 3   Effect of AMP (0.5mM) on PRA Synthesis in Extracts
of Human Lymphocytes.

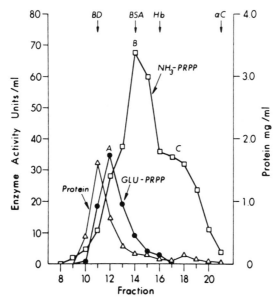

Fig. 4   Elution Profile of Glutamine PRPP Amidotransferase
and Ammonia PRPP Aminotransferase Activity from a G-100
Sephadex Column.

The possibility that two enzyme activities could catalyze PRA synthesis from PRPP was explored. The elution profile of a partially purified enzyme preparation from a G-100 Sephadex column is illustrated on Figure 4. It can be seen that two enzyme activities catalyze PRA synthesis from PRPP, and that each enzyme activity utilizes a different nitrogenous substrate. Peak A, corresponding to an approximate molecular weight of 95,000 daltons, represents maximal glutamine PRPP amidotransferase activity, while peak B (approximate molecular weight of 67,000 daltons) corresponds to maximal PRPP aminotransferase activity. The source of the enzyme preparation of this experiment was Burkitt lymphoma cells maintained in permanent culture.

SUMMARY

The regulation of the early steps of purine biosynthesis of Lesch-Nyhan lymphocytes maintained in culture was compared with that of an established line of normal lymphocytes.

It was found that the rate of FGAR synthesis of Lesch-Nyhan lymphocytes in culture was stimulated 9 fold by the addition of glutamine or ammonia to the incubation medium, while the rate of FGAR synthesis was increased only 4 to 5 fold by these substrates in normal lymphocytes. Maximal rates of FGAR synthesis of these two cell lines as determined in this study, however, were similar, although PRPP concentrations of the Lesch-Nyhan lymphocytes exceeded that of the normal lymphocytes almost 4 fold.

In intact lymphocytes in culture, adenine inhibited the early steps of purine biosynthesis to the same degree in the mutant cell line as it did in the normal cell line. Hypoxanthine inhibited FGAR synthesis of the HGRPT deficient Lesch-Nyhan lymphocytes only in concentrations 30-40 times higher than in normal lymphocytes.

In cell free preparations of HGPRT^{+} lymphocytes it was found that two enzyme activities catalyzed PRA synthesis from PRPP: glutamine PRPP amidotransferase uses PRPP and glutamine as substrates, and ammonia PRPP aminotransferase uses ammonia and PRPP for PRA synthesis.

Comparison of the regulation of PRA synthesis in cell free preparations of Lesch-Nyhan lymphocytes with that of normal lymphocytes showed that the response of these two enzyme activities to inhibition by AMP and GMP of the HGRPT deficient Lesch-Nyhan lymphocytes was similar to that of the normal lymphocyte cell line.

Although in this Lesch-Nyhan lymphocyte cell line the deficiency in HGRPT and the concomitant increase in PRPP concentration was expressed, maximal rates of the early steps of de novo purine biosynthesis were not significantly higher than in the control cell line. The regulation of the rate of purine biosynthesis, therefore, may depend in part on as yet unknown factors of the cellular environment.

REFERENCES

1. Reem, G.H. (1972) J.Clin.Invest. 51, 1058-1062.
2. Reem, G.H. and Friend, C. (1969) Biochim. et Biophys. Acta 171, 58-66.
3. Reem, G.H. and Friend, C. (1967) Science 157, 1203-1204.
4. Reem, G.H. and Friend, C. (1968) J.Clin.Invest. 47, 83a.
5. Reem, G.H. (1968) J.Biol.Chem. 243, 5695-5701.

BIOCHEMICAL CHARACTERISTICS OF 8-AZAGUANINE RESISTANT HUMAN
LYMPHOBLAST MUTANTS SELECTED IN VITRO

G. Nuki, M.B. MRCP, J. Lever, Ph.D., and J. Edwin
Seegmiller, M. D.

Department of Medicine, University of California

San Diego, LaJolla, California 92037, U.S.A.

INTRODUCTION

Much of recent progress in our understanding of the regulation
of purine metabolism in man has resulted from in vitro studies
on cells from patients with inborn errors of metabolism (1, 2).

Advances in tissue culture technology now make it feasible to
induce and select mutations in diploid mammalian cells and so
promise the possibility of extending to human tissue some of the
techniques that have been so brilliantly exploited in prokaryotic
organisms to elucidate molecular mechanisms underlying the control
of gene action (3). Increasing awareness of the extent of genetic
heterogeneity underlying human genetic disease (4,5) also points to
the desirability of establishing model in vitro systems where
phenotypic abnormalities can be studied in a uniform genetic background.

Mammalian cell mutants resistant to the purine analogs
8-azaguanine or 6-thioguanine have been selected in vitro with or
without mutagenesis by a number of workers in mouse L cells (6,7),
Chinese hamster lung cells (8,9), human fibroblasts (10, 11) and
human lymphoblasts (12). Long-term human lymphoblast cultures
have certain advantages for studies of biochemical genetics and
purine metabolism:

1. They can be started from 5-10 ml of venous blood from
    most people.
2. They grow to high density in suspension culture in simple
    nutritional media.

3.   They can be cloned in soft agar.
4.   They maintain normal morphology, karyotype and metabolism over many years apparently free from the problems of senescence.
5.   Many enzymes of purine metabolism detectable in differentiated human tissues are expressed with high activity.
6.   They are relatively inexpensive to culture making it feasible to produce gram quantities if required for enzyme purification and other biochemical studies.

In this paper we describe some biochemical characteristics of fifteen subclones of a permanent human lymphoblast cell line, all of which were derived from a single mass culture resistant to 8-azaguanine.

Cell Material. The parent cell line (WI-L2) (Fig. 1) which was a gift from Dr. Richard Lerner derives from the spleen of a boy with congenital spherocytic anemia. It grows in Eagles Minimal Essential Medium supplemented with glutamine and 10% fetal calf serum. When grown in suspension culture in sealed flasks in a gyratory incubator maintained at 37° it has a doubling time of 22 hours, and it has remained in permanent culture for over five years without change in growth characteristics or karyotype. It has been well-characterized morphologically, immunologically (13, 14) and from the standpoint of purine metabolism (15).

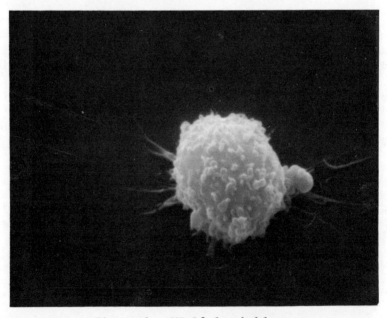

Figure 1. WI-L2 lymphoblast.
Scanning electron micrograph x 4600.

    <u>Selection and Cloning of Resistant Cells</u>. Initially growth curves were performed to determine the minimal inhibitory concentration of 8-azaguanine (Fig. 2).

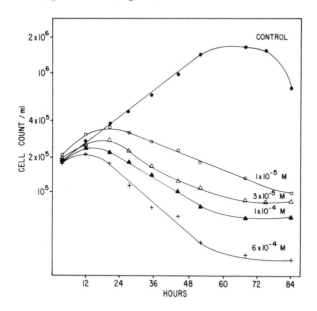

Figure 2. Growth curves for WI-L2 lymphoblasts with increasing concentrations of 8-azaguanine.

    Twenty-five milliliters of an uncloned culture in the logarithmic phase of growth (cell density 6×10^5 cells/ml) were then exposed to 2×10^{-5} M 8-azaguanine in medium. After 48 hours the cells were resuspended in 5 ml of nonselective medium and maintained in increasing volumes of nonselective medium for 10 days prior to reexposure to 2×10^{-5} M 8-azaguanine.

    One month after initial analog exposure there was evidence of partial resistance to 1×10^{-5} 8-azaguanine and full resistance to a concentration of 2×10^{-5} M was achieved after two months.

    Using a modification of the techniques of Coffino, Laskow and Scharff (16) and Lerner, McConahey and Dixon (17), the cells were cloned in soft agar. 2.5×10^3 cells in 3 ml of 0.22% agarose in MEM containing 10 umolar 8-azaguanine were layered into 60 x 15 mm tissue culture dishes containing a confluent feeder layer of WI-38 human embryonic lung fibroblasts covered with a middle layer of 3 ml 0.22% agarose in MEM. MEM plates were placed in a 5% CO_2 incubator at 37°. Using this system clones visible to the naked eye can be detected after one week and picked with a Pasteur pipette under a low-power dissecting microscope after 10-14 days. Cloning efficiency varies from 15-35%.

After picking the individual clones were grown up in nonselec-
tive medium, retested for resistance to 8-azaguanine and recloned
without analog. With a single exception ($Ag^r9Cl_{21}SC_1$) the clones
appeared to be pure as far as can be judged by finding a uniformly
staining population of cells on autoradiography with ^3H hypoxanthine.
All clones were tested regularly for mycoplasma by chemical and
microbiological methods and remained free from contamination. The
growth characteristics of all 8-azaguanine resistant clones were
similar to those of the parent cell line (WI-L2).

Biochemical Characterization. Biochemical characterization of
parent and clonal lines was undertaken on 72-hour cultures at cell
densities of approximately 1×10^6 cells/ml. Results shown are
means of two or more determinations.

Table 1 shows the results of radioassay for HGPRT activity
in cell-free extracts using both .6 mM hypoxanthine $8-^{14}C$
(specific activity 6 mCI/mmole) and .6 mM guanine-$8-^{14}C$ (specific
activity 11 mCI/mmole) as substrates. With minor modifications
the method was that previously described by Fujimoto and Seegmiller
(18) for fibroblasts and Wood, Becker, and Seegmiller (15) for
lymphoblasts. All clones tested showed some degree of HGPRT
deficiency. Three groups can be discerned.

TABLE I

HYPOXANTHINE–GUANINE–PHOSPHORIBOSYLTRANSFERASE ACTIVITY IN CLONED 8-AZAGUANINE RESISTANT HUMAN
LYMPHOBLASTS

| | HGPRTase | | | |
| | ^{14}C Hx | | ^{14}C Gu | |
Cell Line	nmoles/mg protein/hour	% wild type activity	nmoles/mg protein/hour	% wild type activity
Wi-L2 (Control)	327	100	586	100
Ag^r9Cl_1	2	<1	4	<1
$Ag^r9Cl_2SC_1$	18	6	7	1
$Ag^r9Cl_3SC_1$	<1	<1	4	<1
$Ag^r9Cl_{18}SC_3$	3	<1	6	1
$Ag^r9Cl_{35}SC_1$	3	<1	5	<1
$Ag^r9Cl_{16}SC_1$	187	57	353	60
$Ag^r9Cl_{16}SC_2$	190	58	346	59
$Ag^r9Cl_{16}SC_3$	183	56	338	58
$Ag^r9Cl_{20}SC_1$	193	59	301	51
$Ag^r9Cl_{21}SC_2$	104	32	245	42
$Ag^r9Cl_{25}SC_1$	189	58	308	53
$Ag^r9Cl_{26}SC_1$	207	63	360	61
$Ag^r9Cl_{26}SC_3$	48	15	70	12
$Ag^r9Cl_{34}SC_1$	32	10	22	4

A first group of severely deficient clones with 1% or less of
wild-type activity; a second group of partially deficient clones with
enzyme activity ranging from 32-63% of that in the parent strain and
a small third group with intermediate enzyme activity. With the
possible exception of $Ag^r9Cl_2SC_1$ and $Ag^r9Cl_{34}SC_1$ enzyme activity
relative to that in the parent strain was the same using hypoxanthine

or guanine as substrate but guanine resulted in approximately 75% more product formation overall.

Adenine phosphoribosyltransferase activity assayed by an analogous method using 0.6 mM adenine-8-^{14}C (specific activity 4 mCI/mmole) was not significantly altered in any of the HGPRT-deficient clones (Table II). Starch gel electrophoresis using a method only slightly modified from that of Watson et. al. (19) showed no difference in electrophoretic mobility of mutants partially deficient in HGPRT (Fig. 3) but could not be used to examine the clones with severely deficient enzyme activity.

TABLE II

ADENINE-PHOSPHORIBOSYLTRANSFERASE ACTIVITY IN CLONED 8-AZAGUANINE-RESISTANT HUMAN LYMPHOBLASTS.

Cell Line	APRTase	
	nmoles/mg protein/hour	% wild type activity
W1-L2 (Control)	551	100
Ag^{r}9CL$_1$	451	82
Ag^{r}9C1$_2$SC$_1$	492	89
Ag^{r}9C1$_3$SC$_1$	548	100
Ag^{r}9C1$_18$SC$_3$	487	88
Ag^{r}9C1$_35$SC1	540	98
Ag^{r}9C1$_16$SC$_1$	480	87
Ag^{r}9C1$_16$SC$_2$	541	98
Ag^{r}9C1$_16$SC$_3$	503	91
Ag^{r}9C1$_20$SC$_1$	523	95
Ag^{r}9C1$_21$SC$_2$	---	---
Ag^{r}9C1$_25$SC$_1$	495	90
Ag^{r}9C1$_26$SC$_1$	448	81
Ag^{r}9C1$_26$SC$_3$	517	94
Ag^{r}9C1$_34$SC$_1$	521	95

Preliminary studies of heat inactivation in crude undialyzed extracts and mixtures of mutant and wild-type extracts suggest the possibility of a structural gene mutation in two of the partially deficient clonal lines but not in others (Figs. 4 and 5). Altered heat stability has been found in studies of fibroblasts from patients with Lesch-Nyhan Syndrome (4) and partial HGPRT deficiency (20). Mixing experiments rule out the possibility that an inhibitor could be responsible for the lowered HGPRT activity in the deficient clones. The stability of the HGPRT-deficient phenotype over one year in culture further suggests that the mechanism for 8-azaguanine resistance in these cells is mutational rather than epigenetic (21, 22). Growth curves for normal, extremely HGPRT deficient, and partial HGPRT-deficient clones in the presence of 20 uM 8-azaguanine and HAT medium (100 uM hypoxanthine, 0.16 uM aminopterin, 10 uM thymidine) are shown in Fig. 6. As previously noted by DeMars (11) in fibroblasts, cells partially deficient in HGPRT are able to

utilize hypoxanthine sufficiently to survive in HAT medium.
Clearly this places severe limitations on the use of HAT as a
selective medium against HGPRT cells. We have not yet ascertained
whether these HGPRT mutants will show the adaptation to aminopterin
described by Riccardi and Littlefield (23).

The ability of intact cells to incorporate hypoxanthine was
checked by autoradiography using a modification of the method of
Fujimoto and Seegmiller (18) (^{3}H hypoxanthine 10 uM, specific activity 1 ci/umole) and by incorporation of ^{14}C labelled hypoxanthine (59 uM, (specific activity 59mci/mmole) into acid-soluble
nucleotides and acid precipitable nucleic acids (24). Three
autoradiographic patterns were discernible (Fig. 7). Control
WI-L2 cells showed a heavy staining pattern. Clones with severe
HGPRT deficiency showed no more than background staining while
those with more than 30% of wild-type activity in cell-free
extracts showed an intermediate staining pattern approaching that
of true wild type. The two clones with 10-15% of wild-type activity were negative on autoradiography. Representative curves
showing ^{14}C hypoxanthine incorporation into acid-precipitable
nucleic acid and acid-soluble nucleotides following incubation of
intact cells in fresh medium shown in Fig. 8 and whole cell^{14}C
hypoxanthine incorporation at 30 minutes is related to HGPRT levels
in cell-free extracts in Table III. Clones severely deficient
in HGPRT show minimal hypoxanthine incorporation while partially
deficient clones incorporate significant numbers of counts. It is
impossible to be certain whether the negative autoradiography and
low ^{14}C hypoxanthine incorporation exhibited by $Ag^r9Cl_{26}SC_3$ and
$Ag^r9Cl_{34}SCl$ represents a transport defect or merely a relative
insensitivity of the whole cell incorporation studies. Kinetic
studies have yet to be undertaken to determine whether any of the
resistant clones are K_m mutants.

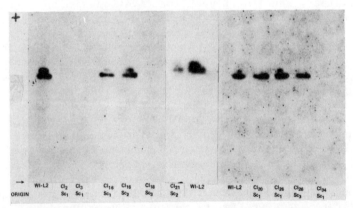

Fig. 3. Starch gel electrophoresis of WI-L2 and 8-azaguanine
resistant clones.

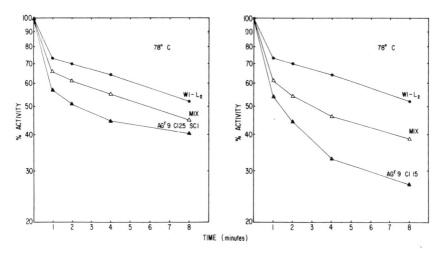

Fig. 4. Heat inactivation curves for crude extracts of WI-L2 lymph-oblasts, two azaguanine resistant clones and mixtures of mutant and wild-type protein. 100% activity was 327 nmoles/mg/hr (WI-L2), 189 nmoles/mg/hr (Ag^{r}9Cl 25 SC 1) and 192 nmoles/mg/hr (AG^{r}9 Cl$_{15}$).

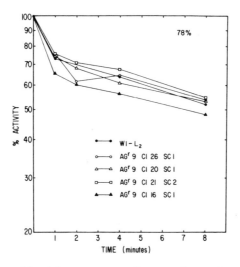

Fig. 5. Heat inactivation curves for crude extracts of WI-L2 lym-phoblasts and four azaguanine resistant clones with partial HGPRT deficiency. 100% activity was 327 nmoles/mg protein/hr (WI-L2), 207 nmoles/mg protein/hr (Ag^{r}9 Cl 26 SC$_1$), 193 nmoles/mg protein/hr (Ag^{r}9 Cl 20 SC$_1$), 106 nmoles/mg protein/hr (Ag^{r}9 Cl 21 SC 1), and 190 nmoles/mg protein/hr (Ag^r- Cl 16 SC 1).

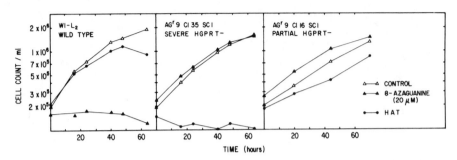

Fig. 6.   Growth curves for wild type (WI-L2), extreme HGPRT⁻ (Ag^{r}9 Cl 35 SC$_1$) and partial HGPRT⁻ (Ag^{r}9 Cl 16 SC$_1$) lymphoblasts in MEM with and without 8-azaguanine (20 µM) and HAT (hypoxanthine 100 µM, Aminopterin 0·16 µM, Thymidine 10 µM).

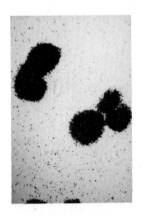
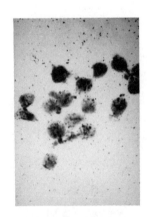

a) WI-L2   b) Ag^{r}9 Cl 1   c) Ag^{r}9 Cl 16 SC 2
    control   severe HGPRT⁻   Partial HGPRT⁻
    very heavy   no staining   Intermediate
    staining   staining

Fig. 7.   Autoradiography with ^{3}H hypoxanthine.

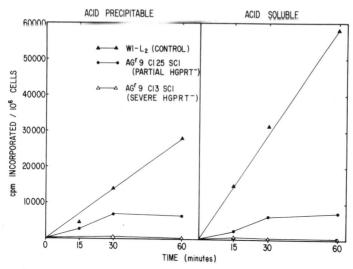

Fig. 8. ^{14}C hypoxanthine incorporation into acid precipitable nucleic acids and acid soluble nucleotides. [1 x 10^6 cells pre-incubated in fresh medium with 5% CO_2 without serum for 15 minutes prior to adding ^{14}C hypoxanthine, regassing and incubating at 37° C.]

TABLE III

WHOLE CELL ^{14}C Hx INCORPORATION RELATED TO HGPRT LEVELS IN CELL EXTRACTS

		^{14}C Hypoxanthine Incorporation			
	HGPRT	Acid Soluble		Acid Precipitable	
Cell Line	(% wild type)	(c/m/10^6 cells)	(% wild type)	(c/m/10^6 cells)	(% wild type)
Wi-L2	100	41388	100	12396	100
Ag^{r}9Cl$_1$	<1	-----	---	-----	---
Ag^{r}9C1$_2$SC1	6	600	1	1024	8
Ag^{r}9C13SL$_1$	<1	269	1	512	4
Ag^{r}9CL$_{18}$SL$_3$	<1	228	<1	372	3
Ag^{r}9CL$_{35}$SL$_1$	<1	613	2	-----	<1
Ag^{r}9CL$_{16}$SC$_1$	57	-----	---	-----	---
Ag^{r}9CL$_{16}$SC$_2$	58	-----	---	-----	---
Ag^{r}9CL$_{16}$SC$_3$	56	11519	28	5867	48
Ag^{r}9C1$_{20}$SC$_1$	59	10738	26	4251	34
Ag^{r}9CL$_{21}$SC$_2$	32	5957	14	4494	36
Ag^{r}9CL$_{25}$SC$_1$	58	6713	16	4287	35
Ag^{r}9CL$_{26}$SL$_1$	63	6032	15	2699	22
Ag^{r}9CL$_{26}$SC$_3$	15	2057	5	839	7
Ag^{r}9CL$_{34}$SC$_1$	10	1813	4	531	4

Intracellular phosphoribosylpyrophosphate concentrations were determined by the conversion of ^{14}C-adenine to ^{14}C-adenylic acid in the presence of highly purified human erythrocyte APRT (25) as previously described (15). Rates of de novo purine synthesis were measured as the rate of incorporation of ^{14}C sodium formate into α-N-formylglycinamide ribonucleotide (FGAR) in the presence of 0.3mM azaserine. The method was similar to that described previously (15) except that incubations were carried out in Eagles MEM containing 10mM/L-glutamine without serum. Results related to HGPRT activity are shown in Table IV. Clones exhibiting severe deficiency of HGPRT show a 4- to 7-fold increase in PRPP and a 2- to 4-fold increase in FGAR accumulation. Those with 30-60% residual HGPRT activity show a 2- to 4-fold increase in PRPP and a lesser, 1- to 2-fold, increase in FGAR accumulated. Measurements of intracellular nucleotide concentrations using the Varian LCS 1000 high-pressure liquid chromatograph and a separation system devised by Dr. David Brenton have shown no significant differences in adenine or guanine nucleotide pools in control and mutant cells (26). These correlations lend some support to the hypothesis (27) that the accelerated rate of de novo purine synthesis exhibited by HGPRT-deficient cells in culture results from increases in intracellular PRPP. The finding that PRPP and FGAR can be further raised by increasing inorganic phosphate concentrations in both normal and mutant cells (Fig. 9), presumably by increasing the activity of PRPP synthetase, provides further evidence for a central regulatory role for PRPP.

TABLE IV

FGAR ACCUMMULATION AND PRPP LEVELS RELATED TO HGPRT DEFICIENCY

Cell Line	HGPRT (% wild type)	PRPP pmoles/10^6 cells	XFold	FGAR c/m/10^6 cells	XFold
Wi-L2	100	10	1.0	35693	1.0
$Ag^{r}9Cl_1$	<1	58	5.8	---	---
$Ag^{r}9Cl_2SC_1$	6	49	4.9	99958	2.8
$Ag^{r}9Cl_3SC_1$	<1	55	5.5	148788	4.2
$Ag^{r}9Cl_{18}SC_3$	<1	71	7.1	130188	3.6
$Ag^{r}9Cl_{35}SC_1$	<1	59	5.9	82115	2.3
$Ag^{r}9Cl_{16}SC_1$	57	26	2.6	44215	1.2
$Ag^{r}9Cl_{16}SC_2$	58	33	3.3	45062	1.3
$Ag^{r}9Cl_{16}SC_3$	56	28	2.8	44308	1.2
$Ag^{r}9Cl_{20}SC_1$	59	37	3.7	66993	1.9
$Ag^{r}9Cl_{21}SC_2$	32	38	3.8	------	---
$Ag^{r}9Cl_{25}SC_1$	58	33	3.3	10024	2.8
$Ag^{r}9Cl_{26}SC_1$	63	32	3.2	74693	2.1
$Ag^{r}9Cl_{26}SC_3$	15	40	4.0	82283	2.3
$Ag^{r}9Cl_{34}SC_1$	10	54	5.4	88735	2.5

A summary of some of the biochemical characteristics of these 8-azaguanine resistant clones is shown in Table V.

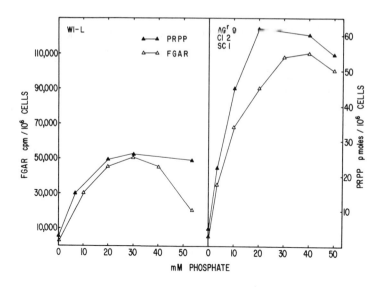

Fig. 9.   Variation of PRPP and FGAR with increasing phosphate
concentration in normal and HGPRT⁻ cells.   [1 x 10⁶ cells/assay
washed with normal saline and incubated in phosphate buffer pH 7.4
for 60 minutes at 37° C prior to preparation for PRPP assay.
FGAR incorporation performed in phosphate buffer with 4 mM glycine,
0.3 mM azaserine, 10 mM glutamine, 5.5 mM glucose.]

TABLE V

Cell Line	Azaguanine Resistance	Growth HAT	Radioautography	Hx Incorporation RNA/DNA	Acid Sol Nucleotides	HGPRTase Hx	Gu	PRPP	FGAR
Wi-L2 CL₂ (Control)	0	+	Dark	100%	100%	100%	100%	1.0	1.0
Ag^{r}9 Cl₁	+	0	Negative	--	--	<1	<1	5.8	--
Ag^{r}9 Cl₂SC₁	+	0	Negative	8	1	6	1	4.9	2.8
Ag^{r}9 Cl₃SC₁	+	0	--	4	<1	<1	<1	5.5	4.2
Ag^{r}9 Cl₁₈SC₃	+	0	Negative	3	<1	<1	1	7.1	3.6
Ag^{r}9 Cl₃₅SC₁	+	0	Negative	<1	2	<1	<1	5.9	2.3
Ag^{r}9 Cl₁₆SC₁	+	+	Partial	--	--	57	60	2.6	1.2
Ag^{r}9 CL₁₆SC₂	+	+	Partial	--	--	58	59	3.3	1.3
Ag^{r}9 CL₁₆SC₃	+	+	Partial	48	28	56	58	2.8	1.2
Ag^{r}9 CL₂₀SC₁	+	+	Partial	34	26	59	51	3.7	1.9
Ag^{r}9 CL₂₁SC₁	+	+	Partial and Negative	36	14	32	42	3.8	--
Ag^{r}9 CL₂₅SC₁	+	+	Partial	35	16	58	53	3.3	2.8
Ag^{r}9 CL₂₆SC₁	+	+	Partial	22	15	63	61	3.2	2.1
Ag^{r}9 CL₂₆SC₃	+	+	Negative	7	5	15	12	4.0	2.3
Ag^{r}9 Cl₁₃₄SC₁	+	+	Negative	4	4	10	4	5.4	2.5

Summary of some biochemical correlates in 8-azaguanine resistant
lymphoblast clones.

SUMMARY AND CONCLUSIONS

1. Fifteen spontaneous 8-azaguanine resistant subclones have been derived from a human lymphoblast cell line (WI-L2).

2. All exhibit varying degrees of HGPRT deficiency and normal APRT activity.

3. Further characterisation by growth in HAT medium, incorporation of radiolabelled hypoxanthing and autoradiography defines three categories of mutants.

4. Intracellular PRPP and rates of de novo purine biosynthesis (FGAR) correlate well with HGPRT deficiency lending support to the hypothesis that elevated PRPP levels are responsible for accelerated de novo purine biosynthesis in HGPRT deficient cells.

5. Starch gel electrophoresis shows no evidence of altered mobility in clones with partial HGPRT deficiency but heat inactivation studies in crude extracts suggest the possibility of more than one structural gene mutation.

6. The biochemical parameters exhibited by these azaguanine resistant human lymphoblast clones appear to be identical to those previously demonstrated in mutant cell lines derived from patients with a wide spectrum of HGPRT deficiency, suggesting that they may be a good model for further studies of the molecular pathology of these diseases.

ACKNOWLEDGEMENTS

This work was supported in part by Grants AM 13622, AM 05646, and GM 17702 of the National Institutes of Health and a grant from the National Genetics Foundation. G.N. is a Merck Foundation International Fellow in Clinical Pharmacology and J.L. is a fellow of the Arthritis Foundation. Miss Inga Jansen gave excellent technical assistance.

REFERENCES

1.   Seegmiller, J. E., Rosenbloom, F. M., and Kelley, W. N., Science 155, 1682 (1967).
2.   Kelley, W. N., Purine and pyrimidine metabolism of cells in culture. In Growth, Nutrition and Metabolism of Cells in Culture, Eds. G. H. Rothblat and V. J. Cristofolo. Academic Press, New York, 1972.

3. Gots, J. S. Regulation of purine and pyrimidine metabolism. In Metabolic Pathways, 3rd Ed., Vol. 5, Metabolic Regulation, Ed. H. J. Vogel, New York, Academic Press, 1971, p. 225.
4. Kelley, W. N., and Meade, J. C., J. Biol. Chem. 246, 2953 (1971).
5. Kirkman, H. N., Enzyme Defects. In Progress in Medical Genetics, Vol. VIII, Eds. A. G. Steinberg and A. G. Bearn, Grune and Stratton, 1972.
6. Liebeman, I., and Ove, P., Proc. Nat. Acad. Sci. USA 45 867 (1959).
7. Littlefield, J. W., Proc. Nat. Acad. Sci. USA 50, 568 (1963).
8. Chu, E. H. Y., Brimer, P., Jacobson, K. B., and Merriam, E. V., Genetics 62, 359 (1969).
9. Gillin, F. D., Roufa, D. J., Beaudet, A. L., and Caskey, C. T., Genetics 72, 239 (1972).
10. Albertini, R. J., and DeMars, R., Science 169, 482 (1970).
11. DeMars, R., and Held, K. R., Humangenetik 16, 87 (1972).
12. Sato, K., Slesinski, R. S., and Littlefield, J. W., Proc. Nat. Acad. Sci. USA 69, 1244 (1972).
13. Levy, J. A., Virolainen, M., and Defendi, V., Cancer 22, 517 (1968).
14. Levy, J. A., Buell, D. N., Creech, C., Hirshaut, Y., and Silverberg, H., J. Nat. Cancer Inst. 46, 647 (1971).
15. Wood, A. W., Becker, M. A., and Seegmiller, J. E., Biochem. Genet., In Press.
16. Coffino, P., Laskow, R., and Scharff, M. D., Science 167 186 (1970).
17. Lerner, R. A., McConaney, P. J., and Dixon, F. J., Science 173, 60 (1971).
18. Fujimoto, W. Y., and Seegmiller, J. E., Proc. Nat. Acad. Sci. USA 65, 577 (1970).
19. Watson, B., Gormley, I. P., Gardiner, S. E., Evans, H. J., and Harris, H. Exp. Cell Res. 75, 401 (1972).
20. Kelley, W. N., Greene, M. L., Rosenbloom, F. M., Henderson, J. F., and Seegmiller, J. E., Ann. Intern. Med. 70, 155 (1969).
21. Harris, M., J. Cell Physiol. 78, 177 (1971).
22. Mezger-Freed, L., Nature (New Biol.) 235, 245 (1972).
23. Riccardi, V. M., and Littlefield, J. W., Exp. Cell Research 74, 417 (1972).
24. Raivio, K. O., and Seegmiller, J. E., Biochim. Biophys. Acta 299, 273 (1972).
25. Greene, M. L., Boyle, J. A., and Seegmiller, J. E., Science 167, 887 (1970).
26. Brenton, D., Nuki, G., Astrin, K., Lever, J., Cruikshank, M., and Seegmiller, J. E., In Preparation.
27. Rosenbloom, F. M., Henderson, J. F., Caldwell, I. C., Kelley, W. N., and Seegmiller, J. E., J. Biol. Chem. 243, 1166 (1968).

Clinical Manifestations
and Genetic Aspects

HYPOXANTHINE - GUANINE PHOSPHORIBOSYL TRANSFERASE DEFICIENCY. OUR EXPERIENCE

B. AMOR - F. DELBARRE - C. AUSCHER - And A. de GERY

Centre de Recherches sur les maladies ostéo-
articulaires U.5 INSERM - Hop. Cochin Paris FRANCE

Thirteen cases of total partial HGPRTase deficiency have been detected at the Rheumatology Clinic of Hopital Cochin (Paris) since Seegmiller, Rosenbloom and Kelley (1) described this enzymatic abnormality.

These cases show very clearly the broad spectra of clinical and biological data already published, as well as particular aspects. Two cases have been suspected only on clinical backgrounds, nine because of unusual responses to drugs (Allopurinol, Thiopurinol and 6 Mercaptopurine) and only 2 by systematic determinations of the enzyme activity on gouty patients.

CASES SUSPECTED ON CLINICAL BACKGROUNDS

Bre... Thierry, first seen in 1967, has a typical Lesh-Nyhan syndrom. Self mutilation appears when he was ten years old in spite of a perfect control of uric acid levels by Allopurinol.
The activity of HGPRTase was very low (<1 n. mole/mg Hb/h. for Hypoxanthine and $<1,5$ for Guanine). But Tab. I shows occasionally an unexplained increase of activity of the enzyme.
Arnold and Kelley (2) have recently described similar variations occuring spontaneously or after an apurinic diet.

date	treatment	Adenine transferase n moles/mg hb/h	Hypoxanthine transferase n moles/mg hb/h	Guanine transferase n moles/mg hb/h
	control m ± sd	22.6 ± 5.07	109.2 ± 12.7	139 ± 21.5
7 . 12 . 67	allopurinol	57	0	0
5 . 1 . 68	nil	59	3.2	6.4
5 . 2 . 68	thiopurinol	59	0	0
5 . 3 . 68	a . m . p	59	0	0
30 . 8 . 72	nil	49	49	30
. . . .	nil	45	2	0.4
16 . 3 . 73	allopurinol	49	1.01	0.7

Tab. I

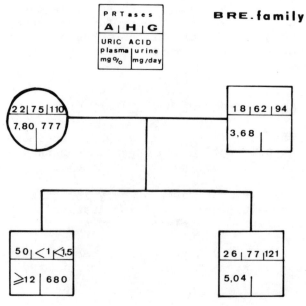

Tab. II

No other deficient subject has been observed in the Bre...
family (Tab. II). It is striking that the Bre... mother, the-
oritically obligatory heterozygote, had normal enzyme activity, not only in erythrocytes but in hair roots as well.
This finding is similar to the inability of Sweetman and Nyhan
to show HGPRTase deficiency in fibroblasts of Lesh-Nyhan mothers (3).
The hypothesis of selection by competition of a normal population of cells seems difficult to be admitted.
In this case, Mercapto Pyrazolo Pyrimidine (Thiopurinol) was
shown to be completely ineffective for reducing both uricemia
and uricuria. On the other hand, Allopurinol produced a decrease of uric acid plasma level but the total purine excretion remained unchanged. This way of responding to treatment
is caracteristic of patients with partial or total HGPRTase
deficiency.

    Did... Martial. This second case, in spite of being a
total deficiency has a milder clinical picture. Unlike the
previous case, this one is able to talk, walk, has never had
sign of self mutilation. His mental retardation is mild ; he
has had only one gouty attack. His mother is clinically normal and erythrocytes HGPRTase activity is normal too.

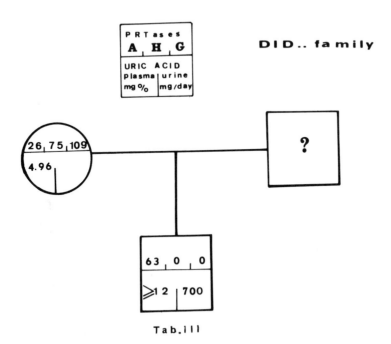

Tab.III

CASES SUSPECTED BECAUSE OF UNUSUAL RESPONSES TO DRUGS

Family Mon... (Tab. V). We discovered this family because
one of its members, a 29 years old man, consulted for a tho-
phaceous gout lasting since he was 17. After treatment by a
usual dose of Allopurinol, we observed that there was no reduction of the total purinuria (Tab. IV) which make us study
the patient and his family looking for HGPRTase deficiency.

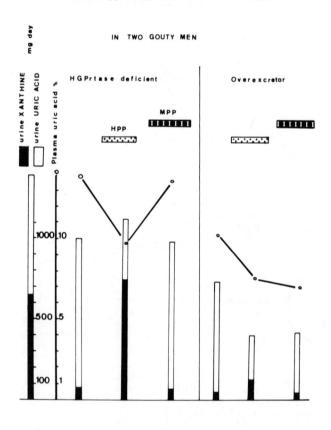

Tab. IV

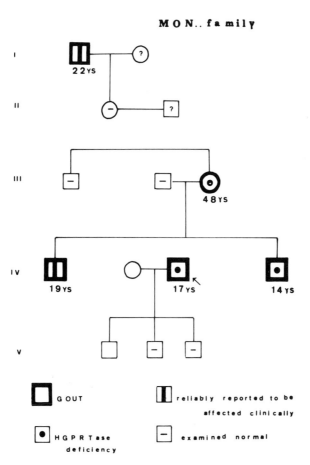

Tab.V

In summary, the important findings are that :
a) The mother of the three gouty brothers is also gouty. She
had her first attack when she was 48, and HGPRTase activity
has been shown to be decreased in her erythrocytes (50% of
normal activity).
b) The mother, in spite of being a partial deficient, behaves
like a total one making Thiopurinol Therapy absolutely ineffective and showing no decrease of total purinuria with Allopurinol.

     Bau... Charles consulted in 1965 when he was 48 years
old for a multitophaceous gout, that he had since he was 20.
The absence of decrease of purinuria and an increase of xan-
thinemia, when treated by Allopurinol, lead us to study the
HGPRTase in 1968 and to find a partial deficiency.

Lug... The deficiency of this gouty patient has been
found in similar conditions. The defect of this subject is
a little unusual.
Red cells Hypoxanthine Transferase is within normal limits
but Guanine Transferase activity is only 30% of normal.
Kelley (4) has mentionned this type of dissociation, but in
none of his patients it was so clear cut. In spite of this
very restricted deficiency, the patient behaves under Allopurinol and Thiopurinol therapy like a total deficient.

Family Mic... (Tab. VII). A patient, Mic... G. (II 6),
negro male from Antillas was admitted to the Hospital in 1958
for a severe tophaceous gout, lasting since he was 28.
By this time, we were looking for the possible action of 6 MP
over the uric acid synthesis. In this patient, we did not observe any action of the drug on uric acid levels in blood or
urine (Tab. VI). This gouty patient had no neurological symp-
tom. He leaves in Africa, working as a teacher.
The important finding in this patient and three other members
of his family is the total HGPRTase deficiency.
II - 8 (43 years old), in spite of being total deficient is
only an over excretor of uric acid.

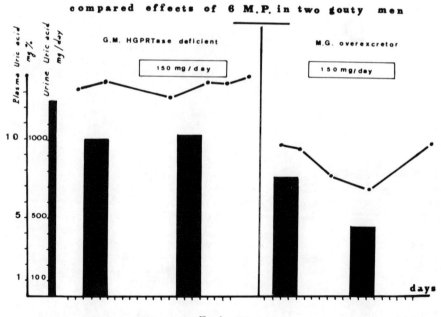

Tab. VI

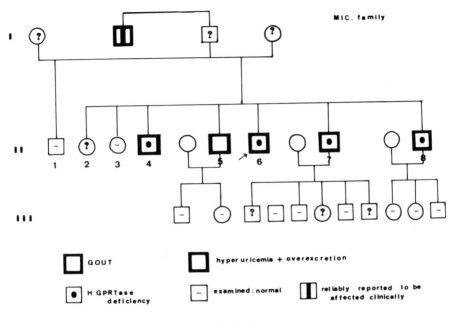

Tab. VII

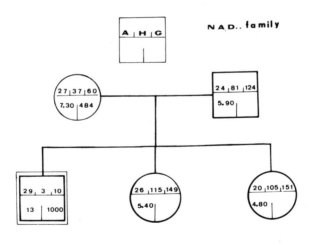

Tab. VIII

CASES DISCOVERED BY SYSTEMATIC DETERMINATION OF THE
ENZYME ACTIVITY

In 1968, a hundred gouty patients were studied searching
for a Purine Transferase activity.

Nad... A. (Tab. VIII) had begun his gout when he was 17
years old. He has developed huge tophie upon his sacro-iliac
joints. His mother had a 50% decrease of the erythrocytes
HGPRTase activity. She has not had, however, gouty attacks
and her serum uric acid levels have been only midly elevated
(7,28 mg %). He has 2 sisters who have normal enzyme levels.

After 1969, we have seen many cases of early gouts and
running in families. But, in spite of that, we have not been
able to find any new case of a Purine Transferase deficiency.
These gouts are associated with kidney insuffisancies which
are also hereditary through an autosomic dominant allele.
These are, most probably, also enzymatic deficiencies but it
would need more research to determine the type of it.

Tab. IX shows the great difference in the genetic transmission of this gene in family Gue...

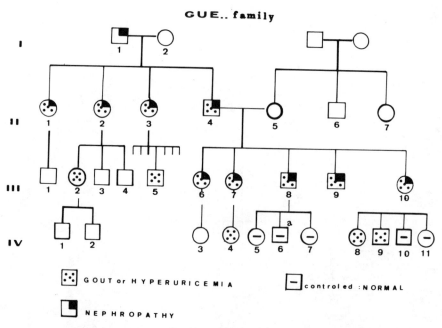

GUE.. family

Tab.IX

REFERENCES

1. SEEGMILLER JE, ROSENBLOOM FM, KELLEY WN.
   Enzyme defect associated with a sex linked humain neuro-
   logical disorder and excessive purine synthesis.
   Science, 152 : 1682-1684, 1967.

2. ARNOLD WJ and KELLEY WN.
   Dietary - Induced Variation of Hypoxanthine - Guanine
   Phosphoribosyl Transferase activity in patients with
   Lesh-Nyhan syndrome.
   J. Clin. Invest., 52 : 970-973, 1973.

3. Sweetman L. and Nyhan WL.
   Further studies of the enzyme composition of mutant cell
   in X linked uric aciduria.
   Arch. Intern. Med., 130 : 214-220, 1972.

4. KELLEY WN, GREENE ML, ROSENBLOOM FM, HENDERSON JF and
   SEEGMILLER JE.
   Hypoxanthine - Guanine phosphoribosyl transferase defi-
   ciency in gout.
   Ann. Intern. Med., 70 : 155-206, 1969

HYPOXANTHINE-GUANINE PHOSPHORIBOSYLTRANSFERASE AND DEUTAN COLOUR

BLINDNESS : THE RELATIVE POSITIONS OF THEIR LOCI ON THE X-CHROMOSOME

B.T. Emmerson, L. Thompson, D.C. Wallace, M. Anne Spence

The University of Queensland Department of Medicine

Institute of Medical Research; University of California

Many studies of X-linked characteristics have suggested that the loci for colour blindness, glucose-6-phosphate dehydrogenase (G-6 PD) and antihaemophilic globulin (AHG) form a fairly close-linked group on the X chromosome, all of which are moderately distant from the locus for Xg antigen (1). Little information is available however concerning the site of the locus for hypoxanthine-guanine phosphoribosyltransferase (HGPRTase) on the X chromosome. This locus has been regarded as being not close to that for G-6 PD because two cross-overs were observed in a single family (2) and no linkage was detected between the HGPRTase locus and Xg antigen (3). Thus, when a patient with a moderate deficiency of HGPRTase, who presented with uric acid calculi and severe gouty arthritis without any neurological signs was found also to be colour blind, members of his family were studied to see if useful information could be obtained concerning the relative positions of the loci for HGPRTase and colour blindness.

The defect of colour vision was tested by the Nagel anomalo-scope in several affected members of the family, and yielded values characteristic of deutan colour blindness. It was found possible to classify members of this family as having colour blindness or normal vision with the use of Ishihara plates and these were used in the field work. HGPRTase activity was measured in erythrocyte lysates by the standard technique of incubation of ^{14}C labelled purine base with lysate and PRPP and measurement of the rate of conversion of purine base to nucleotide (4). Occasionally, when subjects lived in distant regions of Australia or in other countries, finger prick blood spotted on to filter paper was sent through the post (5). The most informative portion of the family consisted of one large sibship of nine males and two females with

their descendants. Seven of the men could be classified by direct
testing and in the eighth, the HGPRTase/colour blindness status
could be deduced from the fact that grandsons showed colour blindness and a daughter showed an intermediate level of erythrocyte
HGPRTase activity (4). One of the two females probably carried
colour blindness because one grandson was colour blind, but no
evidence that she carried the HGPRTase allele was found in that
four grandchildren had normal HGPRTase activity. The other female
of this sibship carried the alleles for both HGPRTase deficiency
and colour blindness. Nineteen descendants of this sibship were
thus able to be classified as either recombinants or nonrecombinants
for HGPRTase and colour blindness, thereby allowing the extent of
recombination of the genes for these characteristics to be
assessed.

The analysis was carried out by a computer programme to assess
linkage data which was similar to the Nulink programme which had
been developed by Renwick and Bolling (6). The lod scores and
relative likelihood values of the various recombination fractions
are shown in the accompanying table and do not demonstrate any
linkage between the loci for colour blindness and HGPRTase
deficiency.

TABLE 1.

Relative Likelihoods of Various Recombination Fractions between
the Loci for HGPRTase and Deutan Colour Blindness

Recombination fraction θ	0	0.1	0.2	0.3	0.4	0.5
Lod score	$-\infty$	-3.274	-1.228	-0.385	-0.067	0

An alternative estimate of the probability of recombination
may also be obtained from the ratio of observed recombinants to
possible recombinants. Four females in this pedigree were doubly
heterozygous because their fathers were both colour blind and
HGPRTase deficient. Together, they had 10 sons. Of these sons,
four had neither colour blindness nor HGPRTase deficiency, two had
colour blindness alone, two had HGPRTase deficiency alone, and two
had both colour blindness and HGPRTase deficiency. Thus four of
these sons were recombinants, an incidence which is not significantly
different from a chance distribution. Thus, the best recombination
fraction which might have been expected would have been close to
0.4, which again would not suggest that the HGPRTase and deutan
colour blindness loci were close together on the X chromosome.

It has been suggested that some of the double heterozygotes
for HGPRTase deficiency and colour blindness should show colour
blindness if selection were occurring against HGPRTase-deficient

cells in double heterozygotes (7). In the present pedigree, no
female was shown to have colour blindness and, on the surface, this
could be taken as evidence against the existence of selection
against HGPRTase deficient cells. However, the HGPRTase deficiency
in this family was only of moderate degree (hemizygotes showed
enzyme activities in erythrocyte lysates between 10 and 20% of
normal and exhibited no neurological signs), so that in this
particular pedigree, where the deficiency was relatively mild, it
would not be surprising if selection did not operate against such
moderately deficient cells. In this particular pedigree, too, a
range of levels of erythrocyte HGPRTase activity intermediate
between normal and the deficient level was found in a number of the
heterozygotes. It is therefore quite possible that selection might
well operate against HGPRTase-deficient cells in those pedigrees
in which the HGPRTase deficiency was gross.

REFERENCES

1. RACE, R.R. and SANGER, R. (1968). "Blood Groups in Man" 5th
   Edition. Oxford : Blackwell pp. 537-556.

2. NYHAN, W.L., BAKAY, B., CONNOR, J.D., MARKS, J.F. and
   KEELE, D.K. (1970). Hemizygous expression of glucose-6-
   phosphate dehydrogenase in erythrocytes of heterozygotes
   for the Lesch-Nyhan syndrome. Proc. Natl. Acad. Sci. USA
   65:214-218.

3. GREENE, M.L., NYHAN, W.L. and SEEGMILLER, J.E. (1970). Hypoxanthine-guanine phosphoribosyltransferase deficiency and Xg
   blood group. Am. J. Hum. Genet. 22:50-54.

4. EMMERSON, B.T., THOMPSON, C.J. and WALLACE, D.C. (1972).
   Partial deficiency of hypoxanthine-guanine phosphoribosyltransferase : Intermediate enzyme deficiency in heterozygote
   red cells. Ann. Intern. Med. 76:285-287.

5. FUJIMOTO, W.J., GREENE, M.L. and SEEGMILLER, J.E. (1968). X-
   linked uric aciduria with neurological disease and self-
   mutilation : Diagnostic test for the enzyme defect.
   J. Pediatr. 73:920-922.

6. RENWICK, J.H. and BOLLING (1967). A program-complex for
   encoding, analysing card storing human linkage data in
   Crow, J.F. and Neele, J.V. (Eds.). Proceedings of the Third
   International Congress of Human Genetics. Baltimore,
   Johns Hopkins, pp. 497-506.

7. DeMARS, R. (1971). Genetic studies of HG-PRT deficiency and
   the Lesch-Nyhan syndrome with cultured human cells. Fed.
   Proc. 30:944-955.

Purine Metabolism and Erythrocyte PRPP Content in Heterozygotes for HGPRT Deficiency

URATE METABOLISM IN HETEROZYGOTES FOR HGPRTase DEFICIENCY

B.T. EMMERSON

University of Queensland Department of Medicine

Princess Alexandra Hospital, Brisbane, Australia

One of the unsolved problems with regard to HGPRTase deficiency has been the occult nature of the abnormality in the heterozygote. Many heterozygotes have shown erythrocyte HGPRTase activity within the normal range (1), although our studies have suggested that this finding is chiefly found in heterozygotes for the severe HGPRTase deficiency which is manifested as the Lesch-Nyhan syndrome. In the less severe HGPRTase deficiency (which is manifested as urate overproduction with minimal neurological signs) our findings have shown a wide range of HGPRTase activities in heterozygotes from as low as 22% of normal up to completely normal values (2). Likewise, the serum urate concentrations in these heterozygotes have most often been found to be normal (1). Some years ago, however, we demonstrated abnormalities of urate metabolism in three heterozygotes with normal serum urate concentrations (3) and we have now extended these studies to a total of nine heterozygotes, all of whom, in one way or another, have demonstrated some abnormality of urate metabolism.

Urate metabolism was defined by a technique which has been evolved by many investigators over the years. This involved the determination of urate pool and turnover by isotope dilution following the intravenous injection of ^{15}N labelled urate, with the simultaneous measurement of the percent incorporation of glycine labelled with an alternative isotopic label (^{14}C) and determination of its percent incorporation into urinary, and also into produced urate, within a 7 day period. (Methods detailed in "The Effect of Weight Reduction on Urate Metabolism").

The results of these studies are shown in Figure 1, although data is as yet incomplete in one subject studied. Seven of the

287

FIGURE 1

Subject	Age (years)	Weight (kg)	S. urate (mg/100 ml)	Urinary urate Excretion (mg/day)	Urate Pool (mg)	Urate turnover (pools/day)	Urate production (mg/day)	Administered urate (%)	Administered glycine (%)	Urate clearance ml/min/1.73m²	Creatinine clearance ml/min/1.73m²
B.W.	38	39.8	4.4	367	535	0.85	455	75.1 (6 days)	0.50 (6 days)	7.5	63.7
R.W.	66	61.7	4.6	348	881	0.63	552	66.4	0.38	4.2	71.2
D.W.	42	54.4	9.1	499	1699	0.58	990	54.7	0.64	6.5	64.0
I.J.	60	69.5	7.5	773	1811	0.68	1232	65.6	0.77	7.1	98
R.L.	55	56.7	4.9	761	868	0.98	853	92.4	1.27	20.1	123.3
L.B.	78	57	5.1	400	834	0.64	535	73.4	0.46	9.6	77.3
C.B.	20	66.7	4.8	586	1026	0.74	763	82.7	0.45	10.7	117
V.B.	41	58.4	4.3	665	734	1.20	884	90.6	0.42	13.7	130
M.B.	43	56.8	4.4	631					0.49	11.8	135

nine subjects had normal serum urate concentrations and all were asymptomatic except for D.W. who had suffered from gouty arthritis for many years. She also had a moderate degree of renal insufficiency and her gout had first occurred following toxaemia of pregnancy, so that the aetiology of her gout is complicated.

As regards urinary urate excretion, only four had urinary urate excretions above the upper limit of normal (600 mg/24 hours) (4). The urate pool, however, whether viewed in absolute terms or in relation to body weight, was clearly abnormal in only two patients and these were the two who were hyperuricaemic. There was considerable variation in the rate of turnover of the urate pool, however, varying between 0.58 to 1.2 pools/24 hours. This latter value is unusually high. Urate production, on the other hand, expressed as mg/kg/day, exceeded the upper limit of normal of 10 mg/kg/day (5) in all but two patients and, when expressed in relation to surface area and taking the normal range as 343 ± 36 mg/square metre/day (6), was clearly abnormal in three of the subjects studied.

The per cent of produced urate which was excreted in the urine was usually within the normal range. The per cent of labelled urate which was excreted in the urine in seven days exceeded the upper limit of normal of 82% (7) in three subjects, and usually agreed well with the per cent of produced urate excreted in the urine.

The per cent administered glycine incorporated into urinary urate in 7 days varied from 0.38% to 1.27%. The usually accepted upper limit of normal for glycine incorporation into urinary urate is approximately 0.3% (7), so that all of these patients incorporate a greater than normal proportion of glycine into urate. The abnormality is emphasized further when the per cent glycine incorporation is corrected to give an estimate of the per cent glycine incorporated into produced urate in this 7 day period. The urate clearance also varied widely, ranging between 4.2 and 20.1 ml/min/1.73m^2 and exceeded 10 ml/min in half of the patients. The creatinine clearances were generally within the normal range except in family W. No explanation can be offered for the reduction in the creatinine clearances in this family except for D.W., who did present other evidence of renal insufficiency.

Thus, each of these nine heterozygotes from four families with varying degrees of HGPRTase deficiency, has demonstrated some abnormality of urate metabolism. We therefore looked to see whether the abnormalities of urate metabolism could be correlated with other measurable factors in these heterozygotes, and took the per cent incorporation of labelled glycine into urinary urate in 7 days as the simplest index of urate production. However, no significant correlation was demonstrated between this parameter and erythrocyte HGPRTase activity expressed as per cent normal. Nor was there any significant correlation between the per cent glycine incorporation into urate and the erythrocyte PRPP concentration.

The presence of both HGPRTase deficient and HGPRTase normal cells has been clearly shown by autoradiographic studies in cultures of fibroblasts from heterozygotes for HGPRTase deficiency (8, 9). It would appear from our findings of the presence of an abnormality of urate metabolism in such heterozygotes that those cells in the heterozygotes which are deficient in HGPRTase are able to induce measurable changes in urate metabolism.

REFERENCES

1. KELLEY, W.N., GREENE, M.L., ROSENBLOOM, F.M., HENDERSON, J.F. and SEEGMILLER, J.E. (1969). Hypoxanthine-guanine phosphoribo-syltransferase deficiency in gout. Ann. Intern. Med. 70:155-206.

2. EMMERSON, B.T., THOMPSON, C.J. and WALLACE, D.C. (1972). Partial deficiency of hypoxanthine-guanine phosphoribosyltransferase: Intermediate enzyme deficiency in heterozygote red cells. Ann. Intern. Med. 76:285-287.

3. EMMERSON, B.T. and WYNGAARDEN, J.B. (1969). Purine metabolism in heterozygous carriers of hypoxanthine-guanine phosphoribo-syltransferase deficiency. Science 166:1533-1535.

4. SEEGMILLER, J.E., GRAYZEL, A.I., LASTER, L. and LIDDLE, L. (1961). Uric acid production in gout. J. Clin. Invest. 40:1304-1314.

5. JAKOVCIC, S. and SORENSEN, L.B. (1967). Studies of uric acid metabolism in glycogen storage disease associated with gouty arthritis. Arthritis Rheum. 10:129-134.

6. RIESELBACH, R.E., SORENSEN, L.B., SHELP, W.D. and STEELÉ, T.H. (1970). Diminished renal urate excretion per nephron as a basis for primary gout. Ann. Intern. Med. 73:359-366.

7. EISEN, A.Z. and SEEGMILLER, J.E. (1961). Uric acid metabolism in psoriasis. J. Clin. Invest. 40:1486-1494.

8. ROSENBLOOM, F.M., KELLEY, W.N., HENDERSON, J.F. and SEEGMILLER, J.E. (1967). Lyon hypothesis and X-linked disease. Lancet 2:305-306.

9. MIGEON, B.R., KALOUSTIAN, V.M.D., NYHAN, W.L., YOUNG, W.J. and CHILDS, B. (1968). X-linked hypoxanthine-guanine phosphoribosyltransferase deficiency : heterozygote has two clonal populations. Science 160:425-427.

ERYTHROCYTE PRPP CONCENTRATIONS IN HETEROZYGOTES FOR HGPRTase

DEFICIENCY

R.B. GORDON: L. THOMPSON: B.T. EMMERSON

University of Queensland Department of Medicine

Princess Alexandra Hospital, Brisbane, Australia

In the last few years, phosphoribosylpyrophosphate (PRPP) has been viewed as a most important intermediate metabolite and studies have indicated that its concentration and rate of synthesis may play a critical role in the regulation of purine biosynthesis *de novo*. Hershko et al (1) found increased *in vitro* formation of PRPP in the erythrocytes of some gouty subjects. They further suggested that the increase in the rate of purine uptake found in these patients was attributable to enhanced PRPP formation. However, other workers (2, 3) failed to find elevated PRPP levels in erythrocytes of patients with gout who had normal hypoxanthine-guanine phosphoribosyltransferase (HGPRTase) activity in their erythrocytes. On the other hand, patients with either severe or moderate degrees of deficiency of the enzyme HGPRTase regularly demonstrated elevated erythrocyte PRPP concentrations. It was suggested (2) that in HGPRTase deficient subjects, the elevated erythrocyte PRPP levels reflected the cells diminished utilization of this compound by the HGPRTase catalysed reaction.

Heterozygotes for HGPRTase deficiency have usually shown normal levels of enzyme activity in erythrocyte lysates even when they were clearly mosaics of HGPRTase-normal and HGPRTase-deficient cells as shown by skin fibroblast cultures. Emmerson et al (4) have demonstrated a range of HGPRTase activities between 22% and 60% of normal in erythrocyte lysates from heterozygotes, although some obligate heterozygotes still demonstrated normal HGPRTase activities. It appeared that these intermediate levels of HGPRTase activity were seen chiefly in heterozygotes for the moderate deficiency and not in heterozygotes for the complete Lesch-Nyhan syndrome.

As PRPP concentration had been demonstrated to be abnormal in erythrocytes from HGPRTase-deficient hemizygotes, it was decided to investigate PRPP levels in erythrocytes from heterozygotes for the moderate HGPRTase deficiency. PRPP concentration in erythrocytes would be an additional parameter useful for the understanding of the abnormalities of uric acid production. It may well be that the increased concentrations of PRPP contribute in part to the increased urate production often found in heterozygotes for the moderate HGPRTase deficiency.

The concentration of PRPP in erythrocytes was measured using the procedure of Sperling et al (5) in which phosphoribosyltransferase activity present in the haemolysate is utilized. PRPP synthesis during the assay was inhibited by 2,3-diphosphoglyceric acid. Since some of the heterozygotes studied had low HGPRTase activity, PRPP was assayed using the haemolysate adenine phosphoribosyltransferase activity with radioactive adenine as the second substrate.

The erythrocyte PRPP concentrations for normal females, for heterozygotes for the Lesch-Nyhan syndrome and for the partial HGPRTase deficiency syndrome are shown in Table I. Also shown in this table are the activities of the hypoxanthine and the adenine phosphoribosyltransferases. These heterozygotes are listed in families in order of decreasing severity of neurological impairment found in the affected male members. Erythrocyte PRPP concentrations were normal in heterozygotes for the Lesch-Nyhan syndrome but were elevated in some of the heterozygotes for the moderate HGPRTase deficiency. No correlation was found between PRPP and the serum urate concentration. Fox and Kelley (3) also found no relationship between intracellular erythrocyte PRPP levels and serum urate concentration or urinary uric acid excretion. It appears, however, that the greater the depression of the erythrocyte HGPRTase activity, the greater is the concentration of PRPP found in erythrocytes ($r=-0.92$; $p < .001$). There is also a correlation between PRPP levels and APRTase activity ($r = 0.89$; $p < .001$). When the PRPP concentration was elevated, the activity of APRTase was also increased.

These findings would therefore support the view that the PRPP concentration in erythrocytes is influenced by the activity of the HGPRTase enzyme, the PRPP content becoming elevated due to a decrease in consumption by the HGPRTase reaction. Rubin et al (6) and Greene et al (7) have attributed the increase in APRTase activity to stabilization of the APRTase enzyme by the high PRPP levels present in the erythrocytes of patients with HGPRTase deficiency. Our findings from the heterozygotes with the moderate deficiency would support this idea of substrate stabilization of the enzyme.

TABLE I.

Erythrocyte PRPP levels and phosphoribosyltransferase activities in heterozygotes for HGPRT-deficiency

	HGPRT Activity of Index case NMOLES/HR/MG	PRPP NMOLES/ML	HGPRT Activity NMOLES/HR/MG	APRT Activity NMOLES/HR/MG	Serum Urate MG/100 ML
Normal (mean±S.D.)		11.6+2.8 (18)	94+7.7(24)	20.1+2.5(22)	4.2+1.2
HETEROZYGOTE					
FAMILY I					
B.W.	<0.005	9.6	89.0	21.1	4.7
E.D.		10.1	85.0	19.0	3.9
E.G		6.0	88.0	18.6	5.5
R.B.		8.4	75.0	18.1	6.5
FAMILY II					
R.L.	<0.03	25.2	10.1	31.5	4.1
FAMILY III					
L.C.	15	10.1	71.0	25.5	4.8
C.C		14.4	52.0	26.5	6.6
M.C.		19.4	28.0	28.4	4.3
V.C.		22.0	17.7	28.2	4.7
FAMILY IV					
M.C.	<0.01	15.3	35.5	28.5	7.3

Evidence against this hypothesis of substrate stabilization has come from Kelley (8) who reported that cultured fibroblasts from patients with HGPRTase deficiency consistently have normal activity of the APRTase enzyme, even though such fibroblasts demonstrate increased intracellular concentrations of PRPP. We have not yet studied the APRTase activity and PRPP levels in fibroblasts from the present series of heterozygotes.

As mutants of PRPP synthetase, the enzyme which catalyses the formation of PRPP have been described (9, 10), it was important to establish the absence of such mutants in the heterozygotes for the partial HGPRTase deficiency. The PRPP synthetase activities of erythrocyte lysates from these heterozygotes were found not to be different from the activities of lysates from normal control subjects. In addition, the PRPP generating capacity of erythrocytes from the heterozygotes was normal. It therefore appears unlikely that the high levels of intracellular PRPP found in the erythrocytes of heterozygotes for the moderate deficiency of HGPRTase are due to increased synthesis of PRPP.

The data presented show that in heterozygotes for the moderate HGPRTase deficiency, the levels of PRPP in erythrocytes reflect the activity of the HGPRTase enzyme. When the HGPRTase activity is low the PRPP levels are elevated above the normal. As a consequence of raised PRPP concentration in the erythrocyte, the APRTase enzyme appears to be stabilized and in erythrocyte lysates the activity of this enzyme is increased.

It is suggested that the determination of PRPP concentration in erythrocytes is an additional parameter to be considered in understanding the biochemical basis for abnormal urate metabolism demonstrated in some heterozygotes for the moderate HGPRTase deficiency.

REFERENCES

1. HERSHKO, A., HERSHKO, C. and MAGER, J. (1968). Increased formation of 5-phosphoribosyl-1-pyrophosphate in red blood cells of some gouty patients. Isr. J. Med. Sci. 4:939.

2. GREENE, M.L. and SEEGMILLER, J.E. (1969). Elevated erythrocyte phosphoribosylpyrophosphate in X-linked uric aciduria : Importance of PRPP concentration in the regulation of human purine biosynthesis. J. Clin. Invest. 48:32a.

3. FOX, I.H. and KELLEY, W.N. (1971). Phosphoribosylpyrophosphate in man : Biochemical and clinical significance. Ann. Intern. Med. 74:424-433.

4. EMMERSON, B.T., THOMPSON, C.J. and WALLACE, D.C. (1972).
    Partial deficiency of hypoxanthine-guanine phosphoribosyl-
    transferase : Intermediate enzyme deficiency in heterozygote
    red cells. Ann. Intern. Med. 76:285-287.

5. SPERLING, O., EILAM, G., PERSKY-BROSH, S. and DeVRIES, A.
    (1972). Simpler method for the determination of 5-
    phosphoribosyl-1-pyrophosphate in red blood cells. J. Lab.
    Clin. Med. 79:1021-1026.

6. RUBIN, C.S., BALIS, M.E., PIOMELLI, S., BERMAN, P.H. and
    DAVIES, J. (1969). Elevated AMP pyrophosphorylase activity
    in congenital IMP pyrophosphorylase deficiency (Lesch-Nyhan
    disease. J. Lab. Clin. Med. 74:732-741.

7. GREENE, M.L., BOYLES, J.R. and SEEGMILLER, J.E. (1970).
    Substrate stabilization : Genetically controlled reciprocal
    relationship of two human enzymes. Science 16:887-889.

8. KELLEY, W.N. (1971). Studies on the adenine phosphoribosyl-
    transferase enzyme in human fibroblasts lacking hypoxanthine-
    guanine phosphoribosyltransferase. J. Lab. Clin. Med.
    77:33-38.

9. SPERLING, O., BOER, P., PERSKY-BROSH, S., KANAREK, E. and
    DeVRIES, A. (1972). Altered kinetic property of erythrocyte
    phosphoribosylpyrophosphate synthetase in excessive purine
    production. Rev. Eur. Etud. Clin. Biol. 17:24.

10. BECKER, M.A., MEYER, L.J., WOOD, A.W. and SEEGMILLER, J.E.
    (1972). Gout associated with increased PRPP synthetase
    activity. Meeting of the American Rheumatism Association,
    Abstract No.8.

Mutants of PRPP Synthetase

MUTANT PHOSPHORIBOSYLPYROPHOSPHATE SYNTHETASE IN TWO

GOUTY SIBLINGS WITH EXCESSIVE PURINE PRODUCTION

O. Sperling, S. Persky-Brosh, P. Boer
and A. de Vries

Rogoff-Wellcome Medical Research Institute and
the Metabolic Unit of Department of Medicine D,
Tel-Aviv University Medical School, Beilinson
Hospital, Petah Tikva, Israel

PRPP is a substrate for the enzyme glutamine-PRPP
amidotransferase catalyzing the first rate-limiting step
of purine nucleotide synthesis de novo (1). Evidence
has been obtained that PRPP is an important regulator
of this pathway (2-4). A possible role of increased
PRPP availability in the enhancement of de novo purine
synthesis in primary metabolic gout, a purine overproduction disease, has been suggested by the demonstration of
an increased PRPP turnover in gouty subjects (5) and of
an increased rate of PRPP formation in erythrocytes and
cultured fibroblasts from such patients (6-8).

The enzyme 5-phosphoribosyl-1-pyrophosphate (PRPP)
synthetase (EC 2.7.6.1) catalyzes the synthesis of PRPP
from ribose-5-phosphate (R-5-P) and adenosine 5'triphosphate (ATP) in the presence of magnesium and inorganic
phosphate (9-11). In the present communication we report
studies on a mutant superactive PRPP synthetase in the
erythrocytes of two brothers with excessive purine
production associated with gout and uric acid lithiasis.
In these two patients the serum uric acid reached 13.5
and 13.6 mg percent and the urinary 24 hours uric acid
excretion 2400 mg and 2250 mg, respectively. All other
members of the family examined were clinically and
biochemically normal, except for the mother of the
patients who had hyperuricosuria, 1100 mg per 24 hours,

but a normal serum uric acid, 5.3 mg percent. A previous study on one of the siblings had shown an increased incorporation of purine bases into the nucleotides of his intact erythrocytes in vitro. Subsequently, the two siblings were found to have an increased erythrocyte PRPP content and in vitro generation (12).

The erythrocyte PRPP synthetase in these patients exhibit in hemolysate, at saturating concentration of ribose-5-phosphate and ATP, a hyperbolic response to increasing inorganic phosphate concentration, as contrasted to the sigmoidal response of the enzyme of of normal control subjects (13,14) (Fig. 1).

Normal hemolysate contains several potent inhibitors of PRPP synthetase such as AMP, ADP, GDP and 2,3-DPG. In order to elucidate whether the hyperbolic response of the mutant PRPP synthetase to increasing phosphate concentration in hemolysate reflects an abnormal response to inhibitors, a system devoid of inhibitors was employed. Using stroma-free charcoal-adsorbed hemolysate treated with DEAE-cellulose, the difference in reaction to increasing inorganic phosphate concentration between the mutant enzyme and the normal enzyme disappeared both exhibiting a hyperbolic response (Fig. 2). It was furthermore found that the mutant enzyme had a decreased sensitivity to inhibition by GDP, ADP, 2,3-DPG and AMP (Fig. 3). At 1 mM inorganic phosphate on the average a 45-fold higher concentration of GDP was needed to inhibit the mutant enzyme to the same extent as the normal enzyme. The differences in the sensitivity to inhibition between the mutant and the normal enzyme were less marked with ADP, 2,3-DPG and AMP, the mutant enzyme requiring 5-, 2-, and 1.5-fold respective inhibitor concentrations. At 1 mM inorganic phosphate and with inhibitor concentrations ADP 0.1 mM, GDP 0.01 mM, 2,3-DPG 4 mM, which are close to those prevailing in the erythrocyte (15), the activity of the mutant enzyme as compared to that of the normal enzyme was 3-, 2- and 1.5-fold, respectively. These activity ratios suggest that in the erythrocyte, in which the concentration of ADP is much higher than that of GDP (16), the increased activity of mutant enzyme reflects its resistance to ADP rather than to GDP. Raising the inorganic phosphate concentration up to 8-10 mM increased the activities of both the normal and the mutant enzyme and decreased their

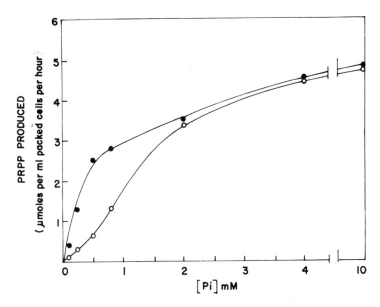

Fig. 1. Activation of normal and mutant PRPP synthetase in hemolysate by inorganic phosphate. o——o normal hemolysate;   ●——● mutant hemolysate

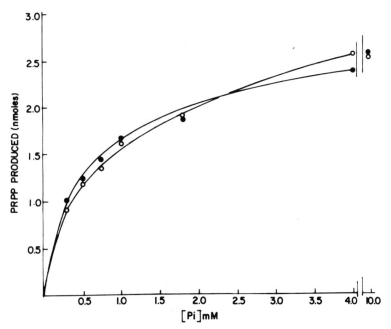

Fig. 2. Activation of partially purified normal and mutant PRPP synthetase by inorganic phosphate. o——o normal enzyme;   ●——● mutant enzyme.

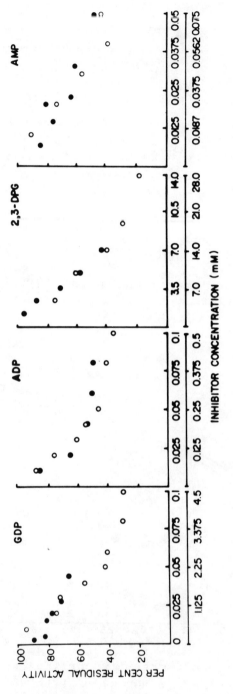

Fig. 3. Inhibition of partially purified normal and mutant PRPP synthetase enzymes by GDP, ADP, 2-3DPG and AMP. Concentration of inorganic phosphate was 1 mM. o control enzyme; ● mutant enzyme.

sensitivity to all inhibitors studied. Furthermore,
at rising inorganic phosphate concentration the difference in the sensitivity of the mutant and normal enzymes
to inhibition decreased, becoming negligible at 8 mM
inorganic phosphate when inhibitor concentrations physiological for the erythrocyte were used. It is noteworthy
that although the physiological erythrocyte inorganic
phosphate concentration is suboptimal for PRPP synthetase
activity, its does render the normal enzyme susceptible
to regulation by various cellular inhibitors. It is
this property in which the mutant enzyme is defective.

The mutant and the normal enzyme were found to have
similar phosphate activation curves (Fig. 2) and similar
substrate saturation curves for R-5-P and for ATP.
Table 1 shows that the partially purified mutant and
normal enzymes did not differ in respect to substrate
and phosphate concentration requirement for half maximal
activity. Both enzymes had identical pH profiles with
maximum activity at pH 7.9, and no difference was found
in the sensitivity of both enzymes to thermal inactivation at 40-65°C. Furthermore, the mutant enzyme could
not be differentiated from the normal enzyme by electrophoresis on cellulose acetate (17).

It thus appears that the mutant PRPP synthetase
enzyme is structurally altered in such a way that only
its regulatory properties but not its catalytic properties are affected. This selective alteration proves
that these two properties are located at different
sites, the enzyme being allosteric. Examples are known
of mutations in bacteria (18,19) and Ehrlich ascites
cells (20) which altered the susceptibility of regulatory enzymes to effector mulecules. An indication for
such a mutation in man has been obtained by Henderson
et al in studies on fibroblasts from two patients with
purine overproduction and gout, showing reduced effectiv-
ness of product inhibition of purine biosynthesis (21).

To our knowledge the present identification of an
erythrocyte PRPP synthetase mutant in the gouty family
studied by us provides the first demonstration in man
of excess activity of a regulatory enzyme causing an
overproduction disease as a direct effect of mutation.

TABLE 1 – SPECIFIC ACTIVITIES IN HEMOLYSATE AND KINETIC
PROPERTIES OF PARTIALLY PURIFIED PRPP SYNTHETASE

Enzyme source	Specific activity in hemolyate (nmoles/ml packed cells/min)	$S_{0.5}$ (mM) of partially purified enzyme		
		R-5-P	ATP	Pi
Normal	78.5 ± 11.74^a	1.2×10^{-2}	7.5×10^{-2}	0.52
Mutant[c]	70.2^b	1.2×10^{-2}	6×10^{-2}	0.48

[a] Mean $\pm$ SD in 12 subjects.
[b] Average of three different determinations
[c] Subject O.G.

References

1. Goldthwaite D.A. J. Biol. Chem. 222:1051, 1956.
2. Fox I.H. and Kelley W.N. Ann. Intern. Med. 74:424, 1971.
3. Kelley W.N., Fox I.H. and Wyngaarden J.B. Clin. Res. 18:457, 1970.
4. Kelley W.N., Fox I. and Wyngaarden J.B. Biochim. Biophys. Acta 215:512, 1970.
5. Jones O.W., Ashton D.M. and Wyngaarden J.B. J. Clin. Invest. 41:1805, 1962.
6. Hershko A., Hershko C. and Mager J. Israel Med. J. 4:939, 1968.
7. Henderson J.F., Rosenbloom F.M., Kelley W.N. and Seegmiller J.E. J. Clin. Invest. 47:1511, 1969.
8. Sperling O., Ophir R. and de Vries A. Eur. J. Clin. Biol. Res. 15:147, 1971.
9. Murray A.W. and Wong P.C.L. Biochem. Biophys. Res. Commun. 29:582, 1967.

10. Wong P.C.L. and Murray A.W. Biochemistry 8:1608, 1969.

11. Switzer R.L. J. Biol. Chem. 244:2854, 1969.

12. Sperling O. Eilam G., Persky-Brosh S. and de Vries A. Biochem. Med. 6:310, 1972.

13. Sperling O., Boer P., Persky-Brosh S., Kanarek E. and de Vries A. Eur. J. Clin. Biol. Res. In press.

14. Hershko A., Razin A. and Mager J. Biochim. Biophys. Acta 184:64, 1969.

15. Barlett G.R. J. Biol. Chem. 234:449, 1959.

16. Mandel P. Folia Haematol. 78:525, 1961-62.

17. Boer P., Sperling O. and de Vries A. In preparation.

18. Moyed H.S. Sympos. Quant. Biol. 26:323, 1961.

19. Ames B.N. and Hartman P.E. Sympos. Quant. Biol. 28:349, 1963.

20. Henderson J.F., Caldwell I.C. and Paterson A.R.P. Cancer Res. 27:1733, 1967.

21. Henderson J.F., Rosenbloom F.M., Kelley W.N. and Seegmiller J.E. J. Clin. Invest. 47:1511, 1968.

INCREASED PP-RIBOSE-P SYNTHETASE ACTIVITY: A GENETIC ABNORMALITY

LEADING TO EXCESSIVE PURINE PRODUCTION AND GOUT*

Michael A. Becker, M. D., Laurence J. Meyer, B. A., Paul

J. Kostel, M. S., and J. Edwin Seegmiller, M. D.

Department of Medicine, University of California

San Diego, LaJolla, California 92037, U.S.A.

In recent years, progress has been made in identifying some of the specific biochemical and genetic factors responsible for the excessive purine synthesis which contributes to the hyperuricemia of a substantial proportion of individuals with gout. Deficiency of glucose-6-phosphatase in Type I glycogen storage disease leads to excessive purine synthesis (1) as do both partial and severe deficiencies of hypoxanthine-guanine phosphoribosyltransferase (HGPRT) (2,3). Another hereditary abnormality of biochemistry has now been described in two families in which purine overproduction and clinical gout are associated with an increased, rather than a decreased, activity of a specific enzyme. The enzyme involved is PP-ribose-P synthetase (4,5).

PP-ribose-P is a sugar phosphate which is synthesized from ATP and ribose-5-phosphate in a reaction requiring magnesium and inorganic phosphate catalyzed by PP-ribose-P synthetase (Figure 1). The PP-ribose-P formed is a substrate in the first and probable rate-limiting reaction of purine synthesis de novo which is catalyzed by PP-ribose-P amidotransferase (PAT). In addition, PP-ribose-P is a substrate in the purine phosphoribosyltransferase reactions which constitute a pathway for the salvage of purine bases.

---

*Abbreviations: PP-ribose-P, 5-phosphoribosyl 1-pyrophosphate;
HGPRT, hypoxanthine-guanine phosphoribosyltransferase;
APRT, adenine phosphoribosyltransferase;
PAT, PP-ribose-P amidotransferase;
FGAR, formyl glycinamide ribonucleotide;
PRA, phosphoribosylamine.

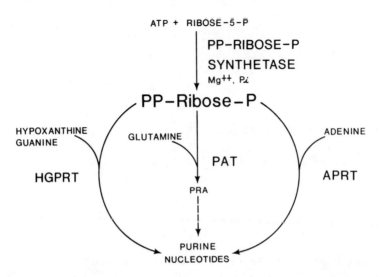

Figure 1. Synthesis of PP-ribose-P and its utilization to form purine nucleotides.

The intracellular concentration of PP-ribose-P is an important regulator of the rate at which purine nucleotides are synthesized (6): altering the intracellular concentration of PP-ribose-P produces a corresponding alteration in the rate of purine synthesis de novo. Thus, in all tissues that have been examined, the striking increases in PP-ribose-P concentration, resulting from deficiency of HGPRT, are associated with marked increases in the rate of purine synthesis de novo. Conversely, depletion of intracellular PP-ribose-P results in decreased purine synthetic rates.

In the course of studying the synthesis of PP-ribose-P in a number of patients with gout and uric acid overproduction, we have found increased PP-ribose-P synthetase activity in erythrocyte lysates and cultured fibroblast extracts from two brothers (T.B. and H.B.) in whom the clinical characteristics of the disease are moderately severe in one brother but not unusually severe in the other and are unassociated with neurologic, mental or hematologic abnormalities (7).

A number of the features of the purine metabolism of these patients are shown in Table 1. In both patients hyperuricemia is moderately severe and each excretes between two and four times the normal quantity of uric acid per day in the urine while receiving a purine-free diet. In both patients the rate of purine synthesis de novo, measured by the incorporation of ^{14}C-glycine into urinary uric acid, is increased 3.0- to 4.0-fold beyond that seen in normal individuals. The urinary uric acid/creatinine ratios in these patients are, as expected, increased above normal. Both APRT and

TABLE 1

Comparison of Purine Metabolism in Normal Individuals and Brothers with Increased PP-Ribose-P
Synthetase Activity*

Group	Plasma Urate	Urinary Uric Acid	Rate of Synthesis	Urinary Uric Acid/Creatinine	Erythrocyte Lysate (nmoles/hr per mg protein)	
					APRT	HGPRT
	mg %	mg/24 hr	%[1-^{14}C] glycine incorporation into urinary uric acid	mean $\pm$ SD		
Normal	< 7.0	413 ± 78 SD	< 0.4	0.34 ± 0.10	16-28	60-106
B. family						
T.B.	10.4	1400-1600	1.4	0.83	19	75
H.B.	9.6	950-1100	1.2	0.65	21	66
C.B.	6.2	890	—	0.55	17	91

* Values given as ranges except as specifically indicated.

HGPRT activities in the patients' red blood cells are normal.

The activity of the enzyme PP-ribose-P synthetase (Table 2), measured in dialyzed hemolysates from these patients, is increased 2.0- to 3.0-fold beyond that seen in normal persons, in a large number of patients with gout (both normal producers and over-producers), and in children with severe HGPRT deficiency and the Lesch-Nyhan syndrome. Two brothers and one sister of the affected patients, as well as the son and wife of patient H.B., have normal enzyme activities. C.B., the asymptomatic and normouricemic 16-year-old daughter of H.B., has activities of PP-ribose-P synthetase comparable to her father and her uncle T.B. In this patient, urinary excretion of uric acid and the urinary uric acid/creatinine ratio are elevated (Table 1). The occurrence of in-creased enzyme activity in a daughter of one of the affected patients suggests that the mode of genetic transmission of the factor responsible for increased PP-ribose-P synthetase activity may be dominant -- either autosomal or X-linked. Since the affected daughter has PP-ribose-P synthetase activities in the range of the affected males, autosomal transmission is more likely although the operation of random X-chromosome inactivation in female heterozygotes (8) is not always demonstrable in erythrocytes as demonstrated by the normal activities of HGPRT in hemolysates from obligate heterozygote mothers of children with severe defi-ciency of HGPRT (9).

The functional significance of the demonstrated increase in PP-ribose-P synthetase activity has been investigated both in red blood cells and in fibroblasts in tissue culture (Table 3). The increase in PP-ribose-P synthetase activity found in hemolysates is also present in extracts of fibroblasts from both patients. The rate of purine synthesis in these patients' fibroblasts, as estimated by the accumulation of ^{14}C-formate into formylglycinamide ribonucleotide (FGAR) in the presence of azaserine, is increased 2.0- to 3.0-fold. Thus, the abnormally high enzyme activity can be demonstrated in a cell that is capable of purine synthesis de novo and the rate of synthesis is increased. In addition, the intracellular concentrations of PP-ribose-P are shown to be in-creased 1.8- to 3.0-fold in the patients' fibroblasts and red blood cells. Of most importance, in the patients' intact fibro-blasts and erythrocytes, the rate of PP-ribose-P, as estimated by measurement of intracellular PP-ribose-P concentration and the incorporation of ^{14}C-adenine into nucleotides and nucleic acids, is increased 1.5- to 1.8-fold. The correlation of increased enzyme activity with increased production of PP-ribose-P by intact cells, provides evidence that the PP-ribose-P synthetase abnormality is of functional significance in the cells. Moreover, the association of increased PP-ribose-P production with purine overproduction strengthens the evidence for a regulatory role of PP-ribose-P in purine synthesis.

TABLE 2

PP-RIBOSE-P SYNTHETASE ACTIVITY IN DIALYZED HEMOLYSATES

Group	Number Studied	Age	Sex	PP-Ribose-P synthetase activity*	
				nmoles/hr/mg protein	
				mean ± 1 SD	range
Normal	28	–	–	65 ± 18	41-98
Gout normal production	16	–	–	69 ± 17	42-92
overproduction	22	–	–	66 ± 15	37-97
HGPRT-deficient	5	–	–	67	53-97
B. family	8	–	–		
T.B.	–	53	M	180	159-188
H.B.	–	43	M	183	160-193
J.B. (brother)	–	50	M	87	79-95
S.B. (brother)	–	51	M	73	–
L.B. (sister)	–	51	F	82	–
B.B. (son of H.B.)	–	15	M	92	89-95
Y.B. (wife of H.B.)	–	40	F	78	75-80
C.B. (daughter of H.B.)	–	16	F	190	185-195

* Activity determined at 32 mM inorganic phosphate (7).

TABLE 3

PP-Ribose-P Metabolism and Purine Synthesis in Brothers with Increased PP-Ribose-P Synthetase Activity.

Cells	Study	T.B.	H.B.
		values relative to means of normal controls (1.0)	
Fibroblasts	PP-ribose-P synthetase activity	3.6	3.8
	Purine synthesis (FGAR accumulation)	2.4	2.5
	PP-ribose-P concentration	2.3	3.0
	PP-ribose-P generation	1.6	--
Erythrocytes	PP-ribose-P concentration	1.8	1.8
	PP-ribose-P generation	1.5	1.8

The increased PP-ribose-P synthetase activity in hemolysates from patients H.B. and T.B. appears to reside in the enzyme itself. Small molecule effectors of the enzyme are removed by dialysis before measurement of the enzyme activity and no activation of normal enzyme or inhibition of mutant enzyme is produced by mixing normal and mutant hemolysates either before or after dialysis. As depicted in Figure 2, inorganic phosphate is a potent activator of the enzyme, and our patients' increased PP-ribose-P synthetase activity is present at all concentrations of added phosphate. This distinguishes the abnormality in PP-ribose-P synthetase activity of our patients from that of a patient described by Drs. Sperling, de Vries, and their colleagues (4), in which increased PP-ribose-P synthetase activity was identifiable only at concentrations of inorganic phosphate below 2 mM. In addition, for that patient's enzyme, evidence has been found for altered sensitivity to nucleotide feedback inhibition, further suggesting a structural alteration in the protein. Despite normal response to nucleotide inhibition, our recent studies indicate that PP-ribose-P synthetase is structurally altered and has increased enzyme activity per molecule in affected members of the B. family.

PP-ribose-P synthetase activity can be detected on cellulose acetate gel strips by means of an activity stain dependent on ATP generation from PP-ribose-P and AMP. Partially purified PP-ribose-

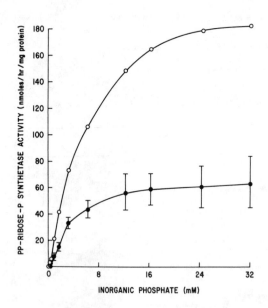

Figure 2.   Phosphate activation of PP-ribose-P synthetase in dialyzed erythrocyte lysates from 15 normal individuals (●——●) and from patient H.B. (O——O).   Brackets indicate ± 1 SD.

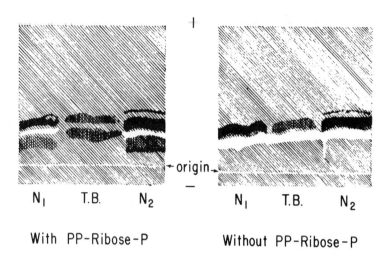

Figure 3. Line drawing of cellulose acetate gel electrophoresis
of partially purified PP-ribose-P synthetase from 2 normal individuals and from patient T.B. The gel developed without PP-ribose-
P in the staining solution represents a control gel. PP-ribose-P
synthetase from patient T.B. migrates further than the enzyme
from controls.

P synthetase from patient T.B. migrates faster in an electrophoretic
field at pH 8.6 than does the enzyme from normal individuals indicating structural alteration in the mutant enzyme (Figure 3).
In addition, identical quantities of a specific rabbit antiserum,
produced in response to 5000-fold purified normal PP-ribose-P
synthetase (10), inactivate about 2.5 times as much enzyme activity
from erythrocytes or fibroblasts of patient T.B. than from similar
cells derived from normal individuals (Figure 4). Thus, the
activity of the B. family enzyme per immunologic unit appears to
be increased in proportion to the increased activity observed in
the patients' cell extracts. Finally, by further immunochemical
studies, the amounts of enzyme detectable in the cells of normal
and B. family members are virtually identical as indicated from
the "equivalence points" shown by the arrows in Figure 5.

SUMMARY

1. In two brothers, increased PP-ribose-P synthetase activity
is associated with purine overproduction and gout.
2. Increased PP-ribose-P synthetase activity is accompanied
in intact cells by an increased PP-ribose-P generation, an elevated

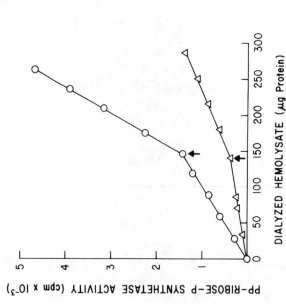

Figure 5. Immune titration of erythrocyte PP-ribose-P synthetase activity from a normal individual ($\triangle$ — $\triangle$) and from patient T.B. ($\bigcirc$ — $\bigcirc$). To a constant volume of immunized rabbit IgG, increasing amounts of hemolysate were added. Enzyme activity was determined after incubation at 37° and 4° each for 30 minutes. Arrows ($\uparrow$) indicate volume of added hemolysate at which unbound enzyme first appears in the incubation mixtures ("equivalence points").

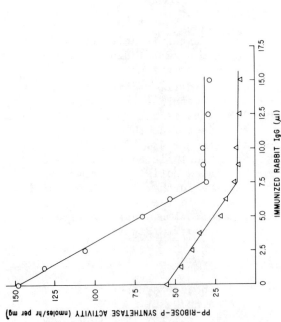

Figure 4. Neutralization of erythrocyte PP-ribose-P synthetase activity by concentrated serum IgG from rabbits immunized with purified normal PP-ribose-P synthetase. To a constant amount of erythrocyte extract, increasing amounts of antiserum were added and after incubation at 37° and 4° each for 30 minutes, enzyme activity was measured. Enzyme from erythrocytes of a normal individual $\triangle$ — $\triangle$; enzyme from erythrocytes of patient T.B. $\bigcirc$ — $\bigcirc$ Residual activity in enzyme-antibody complex appears as flat portion of curves.

intracellular concentration of PP-ribose-P and an increased rate of purine synthesis.

3. These findings suggest that the cause of the patients' gout is increased generation of the regulatory substrate PP-ribose-P resulting from the increased enzyme activity.

4. The molecular basis for the increased enzyme activity appears to be a structural alteration leading to increased enzyme activity per molecule.

5. At least 2 distinct abnormalities in which increased PP-ribose-P synthetase activity is associated with gout have now been described. Increased PP-ribose-P synthetase activity, which appears in the present family to be dominantly inherited, probably with autosomal transmission, represents another genetic disorder leading to purine overproduction and clinical gout. In addition, this abnormality provides the unusual example of a human disease in which an increase rather than a decrease in enzyme activity provides a pathogenetic mechanism for the clinical state.

ACKNOWLEDGEMENTS

This work was supported in part by Grants AM 13622, AM 05646, and GM 17702 of the National Institutes of Health and Grants from the National Genetics Foundation, the National Foundation and the San Diego County Heart Association.

REFERENCES

1. Kelley, W. N., Rosenbloom, F. M., Seegmiller, J. E., and Howell, R. R., J. Pediatr. 72, 488 (1968).
2. Kelley, W. N., Greene, M. L., Rosenbloom, F. M., Henderson, J. F., and Seegmiller, J. E., Ann. Intern. Med. 70, 155 (1969).
3. Seegmiller, J. E., Rosenbloom, F. M., and Kelley, W. N., Science 155, 1682 (1967).
4. Sperling, O., Boer, P., Persky-Brosh, S., Kanarek, E., and de Vries, A., Europ. J. Clin. Biol. Res. 17, 703 (1972).
5. Becker, M.A., Meyer, L. J., Wood, A. W., and Seegmiller, J. E., Science 179, 1123 (1973).
6. Fox, I. H., and Kelley, W. N., Ann. Intern. Med. 74, 424 (1971).
7. Becker, M. A., Meyer, L. J., and Seegmiller, J. E., Amer. J. Med., In Press.
8. Lyon, M. F., Nature (London) 190, 372 (1961).
9. Nyhan, W. L., Bakay, B., Connor, J. D., Marks, J. F., Keele, D. K., Proc. Nat. Acad. Sci, USA 65, 214 (1970).
10. Fox, I. H., and Kelley, W. N., J. Biol. Chem. 246, 5739 (1971).

APRT Deficiency

ADENINE PHOSPHORIBOSYLTRANSFERASE DEFICIENCY: REPORT OF A SECOND FAMILY

I. H. Fox and W. N. Kelley

University of Toronto, Toronto, Canada and Duke University
Medical Center, Durham, North Carolina   27710

Recent advances in the understanding of human purine metabolism have been stimulated by the discovery of specific inborn errors of this pathway in man.   In particular, the demonstration of the deficiency of hypoxanthine-guanine phosphoribosyltransferase (HGPRT) in the Lesch-Nyhan syndrome and in some patients with gout has contributed essential information on the regulation of purine biosynthesis de novo and on the critical role of this reutilization pathway in central nervous system function in man.   The search for other disorders led to the description of a partial deficiency of adenine phosphoribosyltransferase (APRT) in four members in three generations of one family.   Each of the subjects partially deficient in APRT exhibited a normal serum urate concentration and the propositus had a normal excretion of uric acid (Kelley, et al., 1968). We have investigated a second family partially deficient in APRT (Fox and Kelley, in press).

The propositus is a 62 year old black male who had developed podagra in the right great toe at age 46 and began to have recurrent episodes every 2 to 6 months.   When seen at Duke University Medical Center joint aspiration revealed synovial fluid leukocytes containing monosodium urate crystals.   The patient responded to intravenous colchicine.   There was no history of renal calculi, excessive ethanol intake, or gout in the family.   At age 55 the patient had a cerebral hemorrhage and was hypertensive.   Three years later diabetes mellitus was diagnosed.   On physical examination the patient was obese and had a blood pressure of 170/100.   There were no tophi or chronic joint changes.   Laboratory values were as follows: serum urate 13.4 mg/100 ml, urinary uric acid 397 mg/24 hr., uric acid clearance 5.5 ml/min, creatinine clearance 106 ml/min,

cholesterol 198 mg/100 ml, triglycerides 155 mg/100 ml with a normal
lipoprotein pattern.

Erythrocyte APRT deficiency (greater than 3 S.D. below the mean)
was found in 8 subjects in 3 generations from this family (Fig.
1) and in none of the other 347 subjects studied over a 4 year
period. APRT in the propositus was 8.3 ± 3.7 (S.D.) nmoles/mg
protein/hr. In the 4 affected males the mean APRT was 10.4 and in
4 affected females the mean value was 7.6. All family members had
normal HGPRT. In the family pedigree transmission of partial APRT
deficiency occurred from the propositus to all 3 daughters and from
1 affected female to 3 of her 6 children. These findings are most
consistent with the hypothesis that affected individuals are heterozygous for a defect which is inherited in an autosomal manner.
An autosomal locus for the APRT structural gene has been suggested
previously from pedigree analysis, evaluation of human mouse somatic
cell hybrids and an assessment of rare electrophoretic variants.
Hyperuricemia was present in the propositus with APRT deficiency
and in two family members with normal APRT. The other seven APRT
deficient subjects were normouricemic. This illustrates discordance
of APRT deficiency and hyperuricemia in the pedigree.

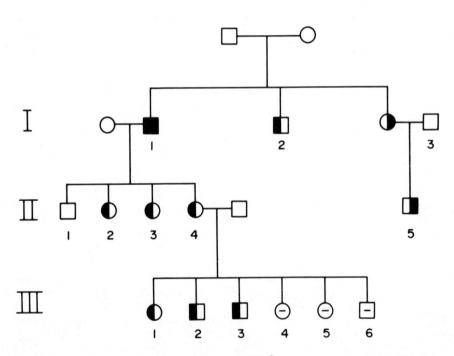

Fig. 1. Pedigree of the L. Family. ▉ ◖ Adenine phosphoribosyltransferase deficiency; ▣ ◗ Hyperuricemia; ▨ ⊖ Examined, normal; ☐ ○
Not examined.

Several properties of the erythrocyte enzyme were studied in order to further characterize the partial deficiency of APRT which was considerably less than 50% of normal values. A mixture of the hemolysate from the propositus with normal hemolysate resulted in the expected intermediate level of activity. This eliminates the possibility of the presence of an inhibitor or absence of an activator of the enzyme in erythrocytes. There was no evidence of increased enzyme lability. APRT from the propositus displayed a normal rate of thermal inactivation at 57°C (Fig. 2). In addition, the apparent half-life of the enzyme in circulating erythrocytes in vivo from the propositus and an affected daughter was similar to the half-life of normal erythrocyte APRT (Fig. 3). The possible explanations for the markedly reduced erythrocyte APRT activity have become complex in the light of recent direct evidence that human erythrocyte APRT normally exists as a trimer composed of three

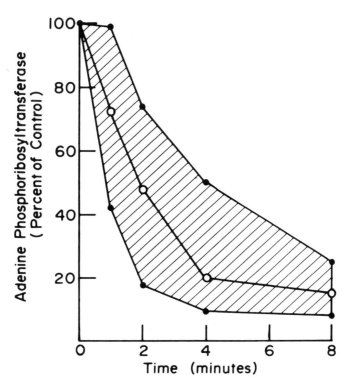

Fig. 2. Heat inactivation of erythrocyte adenine phosphoribosyl-transferase. The incubation was carried out for eight minutes at 57°C. Results are expressed as percent of control activity in unincubated hemolysate. Control, ●—●, (range in 9 subjects); propositus, ○—○.

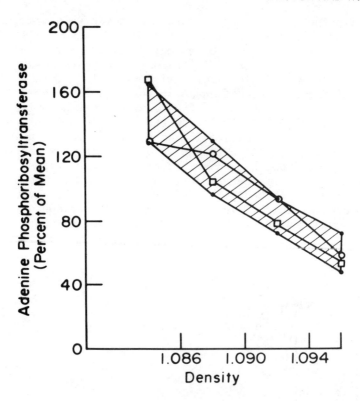

Fig. 3. Effect of erythrocyte density (age) on adenine phosphoribosyltransferase activity. The youngest cells are in the least
dense fraction and the oldest cells are in the most dense fraction.
APRT activity is expressed as per cent of the mean values of all
the fractions. Control, ●—●, (range in 5 subjects); propositus,
O—O; M.L., □—□ .

subunits of identical molecular weight (Thomas, Arnold and Kelley,
1973).

     Further studies were performed to assess the metabolic signif-
icance of APRT deficiency (Table 1). APRT was found to be normal
in leukocytes obtained from two affected members of the family and
in skin fibroblasts cultured from the propositus. Erythrocytes
from the propositus had normal intracellular levels of PP-ribose-P
and ATP. In addition, the oral administration of adenine produced
a marked reduction in the intracellular content of PP-ribose-P in

TABLE 1

STUDIES ON ADENINE PHOSPHORIBOSYLTRANSFERASE DEFICIENCY IN MAN

	Erythrocyte PP-ribose-P (nmoles/ml erythrocytes)	Erythrocyte ATP (μmoles/ml erythrocytes)	Adenine Phosphoribosyltransferase Activity (nmoles/mg/hr)		Urinary Uric Acid (mg/24 hr)	Glycine-^{14}C Incorporation (% of administered dose)
			Leukocytes	Fibroblasts		
Normal (mean±S.D.)	4.4±1.8 (12)	1.22±0.19 (9)	304±87 (21)	211±36 (13)	<590	<0.23
Adenine Phosphoribosyl-transferase Deficiency						
L. L.	5.7	---	234	152	397	0.054
M. L.	---	---	290	---	---	---

Number of subjects indicated in parentheses.

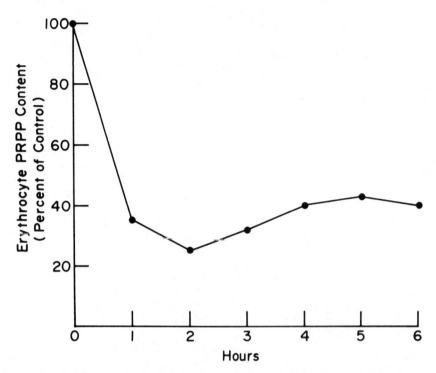

Fig. 4.   Effect of administration of oral adenine on erythrocyte PRPP levels in adenine phosphoribosyltransferase deficiency.

circulating erythrocytes in this patient (Fig. 4). The reduction of intracellular PP-ribose-P was comparable to that observed in subjects with normal or elevated APRT activity (Schulman, et al., 1971). Response to an intravenous fructose infusion was utilized in an attempt to assess adenine nucleotide availability in vivo since fructose produces a rapid catabolism of adenine nucleotides to uric acid (Fox and Kelley, 1972). A rise in the plasma urate and urinary uric acid comparable to that observed in normal subjects occurred following fructose administration to the propositus (Fig. 5). Although the propositus was noted to be hyperuricemic he exhibited a normal urinary excretion of uric acid on a purine free diet as well as a normal rate of purine biosynthesis de novo.

APRT catalyzes the transfer of the ribosylphosphate moiety of PP-ribose-P to the N-9 position of adenine to form AMP (Fig. 6). Utilization of adenine in man seems virtually complete and occurs mainly through the reaction catalyzed by APRT. The daily excretion

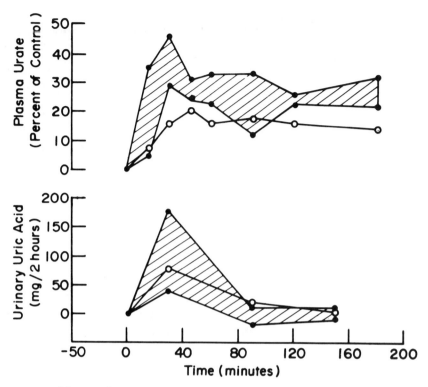

Fig. 5. Effect of intravenous fructose on plasma urate and urinary uric acid excretion. 45 gm of fructose was given intravenously in a 20% solution over a 10 minute period beginning at 0 time. Results are expressed as per cent of control values obtained during the hour before infusion of fructose. Control ●—● (range in 3 subjects); propositus, O—O.

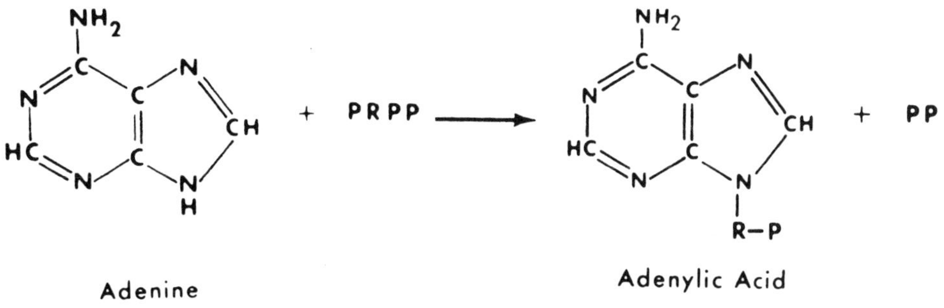

Fig. 6. Reaction catalyzed by adenine phosphoribosyltransferase.

of adenine under normal conditions ranges from 1.1 to 1.7 mg/day.
Despite the theoretically important role of APRT in purine metabolism, the clinical significance of alterations in APRT activity have
been unclear. Elevated APRT has been observed in erythrocytes from
patients with the Lesch-Nyhan syndrome, but the biochemical significance of this finding has not yet been defined. From the present
study it appears that a partial deficiency of erythrocyte APRT
has no effect on purine metabolism. Our studies do not allow one
to conclude that a marked deficiency of the enzyme in all tissues
will be without consequence. In fact a marked or generalized
deficiency of APRT could be lethal.

REFERENCES

Fox, I. H. and Kelley, W. N. 1972. Studies on the mechanism of
    fructose-induced hyperuricemia in man. Metabolism 21: 713-721.

Fox, I. H., Meade, J.C. and Kelley, W. N. Adenine phosphoribosyl-
    transferase deficiency: Report of a second family. Amer. J.
    Med. (in press).

Kelley, W. N., Levy, R. I., Rosenbloom, R. M., Henderson, J. F.
    and Seegmiller, J. E. 1968. Adenine phosphoribosyltransfer-
    ase deficiency: A previously unrecognized genetic defect in
    man. J. Clin. Invest. 47: 2281-2289.

Schulman, J. D., Greene, M. L., Fujimoto, W. Y. and Seegmiller, J.
    E. 1971. Adenine therapy for Lesch-Nyhan syndrome. Ped.
    Res. 5: 77 - 82, 1971.

Thomas, C. B., Arnold, W. J. and Kelley, W. N. 1973. Human
    adenine phosphoribosyltransferase: Purification and proper-
    ties. J. Biol. Chem. 248: 144-150.

ADENINE PHOSPHORIBOSYLTRANSFERASE DEFICIENCY IN A FEMALE

WITH GOUT

B.T. EMMERSON, R.B. GORDON and L. THOMPSON

University of Queensland, Department of Medicine

Princess Alexandra Hospital, Brisbane, Australia

A moderate deficiency of adenine phosphoribosyltransferase (APRTase) has been described in two families (1,2). In the first of these families the deficiency of APRTase extended to four members in three generations. However in these cases, there was no evidence of abnormal urate metabolism. The second report of a moderate APRTase deficiency involved a patient with a long history of gouty arthritis. In this patient, the incorporation of ^{14}c-glycine into urinary uric acid and the urinary excretion of uric acid were normal, and the authors concluded that the moderate deficiency of APRTase was not associated with a derangement of *de novo* purine synthesis. A functionally similar enzyme, hypoxanthine-guanine phosphoribosyltransferase (HGPRTase), also involved in purine metabolism, has been reported to be only partially active in cases of gout in which the subjects produce excessive quantities of uric acid (3).

In the present study a female patient, also having a history of gouty arthritis, has been found to have a moderate deficiency of APRTase. Since the finding of gout in young females is uncommon, the present case afforded another opportunity of studying the possible connection between the clinical character-istics and the deficiency in the APRTase enzyme. Several members of the patient's family have also been studied.

The propositus, a woman aged 24 years, had suffered from recurrent gouty arthritis since the age of 11 years. She also demonstrated considerable, though asymptomatic, renal impairment with a creatinine clearance of approximately one-third normal. Both the propositus and her mother were shown to have reduced

327

TABLE I

Details of urate metabolism in female with gouty arthritis.

MRS. R.K.   24 years   (51 kg; S. area 1.55m^2)

	PATIENT	NORMAL
Mean serum (urate purine free diet) (mg/100ml)	10.2 ± 0.5	4.2 ± 1.2
Mean urate clearance (ml/min/1.73m^2)	1.7	9.2 ± 2.7
Mean creatinine clearance (ml/min/ 1.73m^2)	41	108 ± 16
Miscible urate pool (mg)	1644	1085 ± 190
95% limits of reliability of pool estimate (mg)	1536-1760	
Turnover rate of pool (pools/day)	0.31	0.59 ± 0.09
Urate production (mg/day)	507	620 ± 60
Urinary urate excretion (mg/day)	212 ± 21	275-600
Intravenous urate excreted in 7 days (%) (A)	37.1	70 ± 8
Glycine incorporation into urinary urate in 7 days (% dose) (B)	0.19	0.17 ± 0.07
Glycine incorporation into produced urate in 7 days (% dose) ($\frac{B}{A}$ x 100)	0.51	0.31 ± 0.04

activity of erythrocyte APRTase (45% and 42% respectively of the
mean value from 78 normal control subjects). The decreased
activities were shown not to be due to the presence of a
reversible inhibitor. The male members of the family had
erythrocyte APRTase activities within the normal range. In all
family members studied the specific activity of HGPRTase was
found to be normal. Whether the APRTase activity was normal or
not, a normal concentration of erythrocyte PRPP was found.

Some biochemical studies of the patient's APRTase have
been carried out. Heat inactivation studies according to the
procedure of Henderson et. al.(4), showed the thermostability of
the patient's APRTase to be no different from normal. Partial
purification of APRTase from the patient's erythrocytes was
carried out by the procedure of Rubin et.al.(5), purification being
taken to the 47-70% ammonuim sulphate fraction and resulting in
100-135 fold increase in specific activity. The APRTase enzymes
from two normal controls were also purified in a similar manner.
Kinetic studies demonstrated that with the deficient enzyme the
K_m values for the two substrates, adenine and PRPP, were not
significantly different from the values obtained with the normal
enzyme preparations. Polyacrylamide gel electrophoresis of the
partially purified enzymes revealed no difference in electrophoretic mobilities. Thus, no evidence obtained so far has
suggested any structural alteration of the APRTase obtained from
the propositus, although it is recognised that many structural
gene mutations may occur which might affect enzyme activity
without grossly altering other properties of the protein.

Urate metabolism in the propositus was defined on a purine-
free diet and details are shown in Table 1. Unlike the patient
reported by Kelley et.al(2) there was very little fall in the
serum urate concentration with purine restriction. There was,
however, a considerable reduction in glomerular filtration rate
as reflected by the creatinine clearance, with a considerable
impairment of the urate clearance as well. The miscible urate
pool was considerably increased (normal for this patient
approximately 1000mg). However, the rate of turnover of this
pool was slower than usual so that the daily production of urate
was within the normal range. The urinary excretion of urate was
also abnormally reduced, probably attributable to the degree of
renal impairment, and there was a corresponding reduction in the
percentage of labelled urate which was excreted in the urine
during a seven day period. The incorporation of ^{14}C-glycine into
urinary urate in seven days was within the normal range, although
when this was corrected to give a value for the percentage
incorporation of glycine into the urate which had been produced
in this time, a value was obtained which was probably somewhat
increased above normal.

The first patient described by Kelley et.al.(1) also had an
abnormal lipoprotein pattern consistent with Type II hyperlipoproteinaemia. No other family member, even those with reduced
APRTase deficiency, had Type II hyperlipoproteinaemia. In the
present family the serum cholesterol, serum triglycerides and
lipoprotein pattern were determined and were normal for the
propositus and her mother. However her father, who had normal
APRTase activity, had an elevated concentration of serum trigly-
cerides and showed a raised pre-β band, both characteristic of
Type IV hyperlipoproteinaemia. There is thus no essential
association between APRTase deficiency and a disorder of lipoprotein
metabolism, since they occur independently in the present kindred.

The gouty arthritis in the propositus was atypical in regard
to the age of onset, the sex and the severity of the renal disease.
Thus, the type of gouty arthritis exhibited by the propositus
showed uncommon features and was associated with an unusual pattern
of urate metabolism. The mother of the propositus, although her
APRTase activity in erythrocyte lysates was also decreased,
demonstrated serum and urinary urate concentrations towards the
upper limit of normal and hence showed no definite evidence of
abnormal urate metabolism. However, the father of the propositus
had gout but this was of the type commonly associated with obesity
and regular alcohol consumption, and sometimes with hypertrigly-
ceridaemia (6).

Although the relationship between propositus and father would
suggest a common basic inherited metabolic abnormality, it seems
most likely that the cause of the hyperuricaemia in the propositus
was quite different from that in her father. Viewing this family
alone, there is no definite reason to relate the deficiency of
APRTase to the abnormality resulting in the hyperuricaemia and
gout. However, three families with APRTase deficiency have now
been described and two patients have had gout, so it is less
convincing to dismiss as a chance association.

Kelley et.al.(1) concluded from their genetic analysis of the
first family with the deficiency, that the patients seen were
heterozygotes and that the deficiency was inherited by an autosomal
mechanism. Such a conclusion could apply to the present family.
The son of the propositus however, had a normal APRTase activity
in erythrocytes, thereby demonstrating no evidence that he had
inherited the deficiency. Further genetic studies of this family
are underway.

REFERENCES

1. KELLEY, W. N., Levy, R. I., ROSENBLOOM, F. M. HENDERSON, J. F., and SEEGMILLER, J. E. (1968). Adenine phosphoribosyl-transferase deficiency : A previously undescribed genetic defect in man. J. Clin. Invest. 47 : 2281-2289.

2. KELLEY, W. N., FOX, I. H., and WYNGAARDEN, J. B. (1970). Further evaluation of adenine phosphoribosyltransferase deficiency in man. Occurrence in a patient with gout. Clin. Res. 18, 53.

3. KELLEY, W. N., ROSENBLOOM, F. M., HENDERSON, J. F., and SEEGMILLER, J. E. A specific enzyme defect in gout associated with overproduction of uric acid. Proc. Natl. Acad. Sci. U. S. A. 57, 1735, (1967).

4. HENDERSON, J. F., MILLER, H. R., KELLEY, W. N., ROSENBLOOM, F. M., and SEEGMILLER, J. E. Kinetic studies of mutant human erythrocyte adenine phosphoribosyltransferases. Can. J. Biochem. 46, 703-706 (1968).

5. RUBIN, C. S., DANCIS, J., YIP, L. C., NOWINSKI, R. C., and BALIS, M. E. Purification of IMP : pyrophosphate phosphoribosyltransferases, catalytically incompetent enzymes in Lesch-Nyhan disease. Proc. Nat. Acad. Sci. U. S. A. 68, 1461-1464. (1971).

6. EMMERSON, B. T. and KNOWLES, B. R. Triglyceride concentrations in primary gout and gout of chronic lead nephropathy. Metabolism 20, 721-729. (1971).

GOUT WITH ADENINE PHOSPHORIBOSYL TRANSFERASE DEFICIENCY

F.DELBARRE, C.AUSCHER, B.AMOR, A. de GERY
Institut de Rhumatologie - Unité n° 5 INSERM -ERA 337 CNRS
Hôpital Cochin   27, rue du Fg St Jacques
75014   PARIS   FRANCE

INTRODUCTION

A systematic study of purine phosphoribosyl transferases has led to the discovery of a deficiency of adenine phosphoribosyl transferase (APRT) in a patient with gout and several members of his family, who were either suffering from gout, hyperuricaemia or were normal.

This to our knowledge is the second case of gout associated with APRT deficiency; an observation of this combination has been reported by Kelly et al. (8,9). We have made clinical and bio-chemical examinations of the patient and as many of his family as possible, to learn more about the gout or the abnormal purine metabolism.

MATERIAL AND METHODS

A radio-enzymatic method was used to measure the APRT activity in the red cells. From the enzymatic conversion of 14 C adenine and its radio-active nucleotide in the presence of a red cell lysate and PRPP, the percentage of ribosyl phophorylation that is calculated and expressed as an moles of adenine transformed into its nucleotide per hour, per gramme of protein (3,6). According to Seegmiller (6) and Cartier (3), the mean values ($\pm$ 2SE) are 21 ± 10.2 and 24.12 ± 15.16 n.m/mgHb/h. respectively.

The biochemical measurements were made on 2 subjects with gout whilst in hospital, on a low purine diet with a constant intake of

333

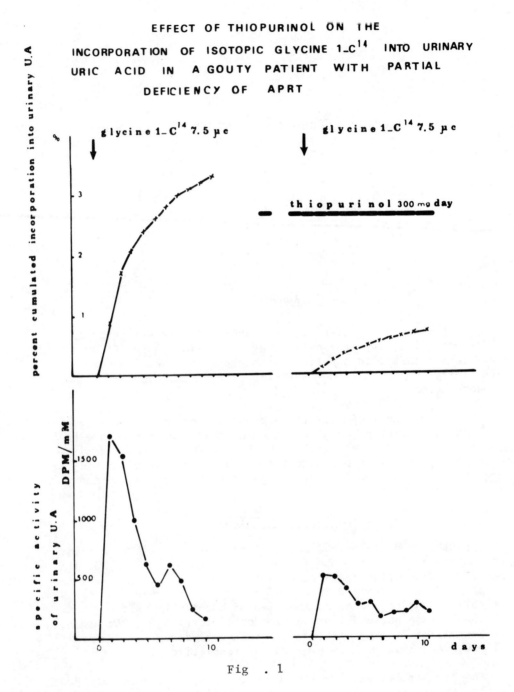

Fig . 1

calories, proteins, lipids and carbohydrates.

The quantity of urinary oxypurines (hypoxanthine + xanthine) were determined on 24 urine collections using a double enzyme technique (xanthine oxidase + uricase) according to Jorgensen and Poulsen (5). The mean normal values were 23.2 ± 13.4 mg/24 hours. The incorporation of 14 C-glycine into the urinary uric acid was measured from the analysis of urine for IO days after the ingestion of 7 µ Ci of I4 C-I-glycine (CEA-Saclay) and expressed as % of the dose administered (2). Normally this incorporation is less than 0.2% in the common forms of gout 0.3 to I% and in gout with hypoxanthine guanine transferase deficiency 2.8% (7).

The biochemical studies of the propositus involved observations on his reactions to hydroxy 4 pyrazolo 3-4 d pyrimidine (Allopurinol) and mercapto 4 pyrazolo 3-4 d pyrimidine (Thiopurinol).

In the family studies the examinations were limited to measurement of the uric acid and oxypurines in the blood and urine, either 24 hour collections or single samples or urine, and of the calculation of the ratio of uric acid to creatinine.

The karyotype was examined in most subjects to see if there was any relation between a chromosome anomaly and the familial deficiency of ARPT.

RESULTS

Monsieur Rene GIR, aged 50 years old, suffered from acute gout since the age of 44 after   left nephrectomy for pyelonephritis and several attacks of renal colic and passive red gravel in his urine.

There had been only three acute attacks of gout during 6 years, without tophi or arthropathy. His blood uric acid was IOO mg./IOOml , and a mean urinary output of uric acid in the urine of 970 mg/24 hours with spontaneous crystaluria, the mean output of xanthine in the urine was 26 mg./24 hours. His renal function was normal (uric acid clearance IO.5 ml., blood creatinine I.04 mg./IOO ml). All other tests including blood lipid concentrations were normal.

The hyperpurine productive character of his gout was confirmed by the 3.3% incorporation of I4 C-glycine into the urinary uric acid and the shape of the output curve (Fig.I). This patient suffered from primary gout (there was no renal failure, obesity, intoxication or polycythemia).

The APRT activity in the red cells was 4 n moles/mg. protein/
hour (mean of three measurements), this is about 15% of the normal
value. These measurements were confirmed by Dr. Seegmiller.

EFFECTS OF INHIBITORS OF PURINE SYNTHESIS

Allopurinol (200 mg/24 h.) reduced the uricaemia to 5.7 mg%
and the uraturia/24 hr. to 310 mg, increased the xanthinuria (hypoxanthine + xanthine) but the level of oxypurine (hypoxanthine +
xanthine + uric acid) excretion in the urine was reduced by 43%.

Thiopurinol (300 mg./24 h.) reduced the uricaemia to 4.66 mg.%
and the uraturia to 290 mg/24 h without increasing the "xanthinuria". It reduced by 76% the incorporation of 14 C-glycine into
the urinary uric acid during a ten day period.

FAMILY STUDY (fig. 2)

This was conducted on 23 persons belonging to four generations
(7 females and 16 males). In 7 males and one female the red cell
APRT was reduced to at least 50%. Only two of the subjects had gout
(the propositus II - 2 and his brother II - 4). Three had urinary
lithiasis (the propositus II - 2, a brother II - 4, and a nephew III -
13). Seven had either biochemical or clinical signs or both of uric

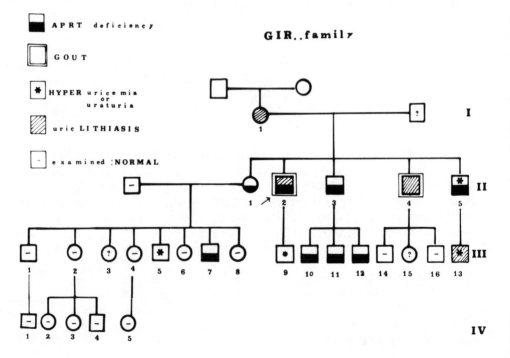

acid hyperproduction. The occurrence of a trisomy 2I (generation
ITI) was found in a systematic study of I3 members of the family
(Professor Lejeune) .   Chromosome anomalies (translocations D∿G)
were found in 8 out of the I3, unrelated to the enzymopathy.

DISCUSSION

This fortuitous observation of a case of gout with a deficiency
of an enzyme concerned in the regulation of uric acid synthesis
by feed back, raises the question whether this deficiency is responsible for the abnormal purine metabolism and uric acid hyper-
synthesis.

If these metabolic disorders are considered to be related,
then the cases of gout observed in this family are different in
their manifestations to those resulting from a deficiency of hypoxanthine guanine phosphoribosyl transferase (4). They are characterised by a slow onset, few clinical signs and especially absence
of tophi and a remarkable effect of allo- and thiopurinol.

The family study showed in some members there was an association between the APRT deficiency and signs of abnormal purine metabolism, and others where the enzyme deficiency was without any
apparent abnormality of purine metabolism. However, this cannot
be the explanation for existence of urinary lithiasis in subject
III - I3 when the level of APRT activity in the red cells was
normal.

As yet there is no firm evidence to suppose the hyperproduction of uric acid in the propositus is the result of deficiency
of APRT and a raised feed back.   Although this may be the mechanism in the case of gout associated with a hypoxanthine guanine
phosphoribosyl transferase deficiency.

SUMMARY

Screening the purine phosphoribosyl transferases in patients
with gout has demonstrated an adenine phosphoribosyl transferase
deficiency in one of them.

Analysis of 23 members of the family showed others with the
enzyme deficiency without abnormal purine metabolism. This gout
is associated with very high uric acid production, and marked
sensitivity of Allopurinol and Thiopurinol, but these two factors
and APRT deficiency, as in gout with a hypoxanthine guanine P.R.
transferase deficiency, could not be established.

REFERENCES

I - Auscher C., Brouilhet H. et Delbarre F.
    Le taux des oxypurines dans le plasma et les urines de sujets
    témoins et goutteux. (Correlation avec le taux d'acide urique)

    Sem. Hop. Paris - I967 - <u>43</u> - II23.

2 - Auscher C., Delbarre F., Roucayrol J.C. et Ingrand I.
    Effets du thiopurinol sur le metabolisme des purine; etude de
    l'incorporation de la glycine I4 C dans les urates.

    IIe table ronde de langue française sur l'exploration fonction-
    nelle par les isotopes radioactifs. (Lausanne mai I969).

3 - Cartier P. et Hamet M.
    les activités purine - phosphoribosyl transferasiques dans les
    globules rouges humaines. Technique de dosage.

    Clin. Chim. I968 - <u>20</u>, I05.

4 - Delbarre F., Cartier P., Auscher C., de Gery A., et Hamet M.
    Gouttes enzymopathiques. I. Gouttes avec déficit en hypoxanthine
    guanine phosphoribosyl transferase.

    La Presse Médicale - I970, <u>78</u>, I6.

5 - Jorgensen S. and Poulsen. H.E.
    Enzymatic determination of hypoxanthine and xanthine in human
    plasma and urine.

    Acta pharmacol. I955, <u>II</u>, 223.

  6- Kelley W N., Rosenbloom F.M., Henderson J.F., and Seegmiller J.E
    A specific enzyme defect in gout associated with overproduction
    of uric acid.

    Proc. Nat. Acad. Sci. I967, <u>57</u>, I735.

7 - Kelley W.N, Greene M.L., Rosenbloom F.M., Henderson J.F. and
    Seegmiller J.E.

    Hypoxanthine, guanine, phosphoribosyl transferase deficiency in
    gout.

    Annals of Internal. Med. I969, <u>70</u>, I55.

8 - Kelley W.N., Levy R., Rosenbloom F., Hendersen J., and
Seegmiller E.
   Adenine phosphoribosyl transferase deficiency : A previously
   undescribed genetic defect in man.

   J. Clin. Inv.   1968, $\underline{47}$, 2281.

9 - Kelley W.N., Fox I.H , and Wingaarden J.B.
   Further évaluation of adenine phosphoribosyl transferase deficiency in man. Occurence in a patient with gout.

   Clin. Res.   1970, $\underline{18}$, 53.

IO - Praeterius E., and Poulsen H.E.
   Enzymatic determination of uric acid.

   Scand. J. Clin. lab. invest.   1953 , $\underline{5}$, 273.

II - Wyngaarden J.B.
   Overproduction of uric as the cause of hyperuricaemia in primary gout.

   J. Clin. Invest.   1957 , $\underline{33}$, 1508.

Xanthinuria

XANTHINURIA IN A LARGE KINDRED

D. M. Wilson and H. R. Tapia

Mayo Clinic and Mayo Foundation

Rochester, Minnesota   USA

Xanthinuria is a rare hereditary disorder characterized by a gross deficiency of xanthine oxidase activity in tissues with a resultant decrease in urinary uric acid excretion and a concomitant increase in the excretion of xanthine and hypoxanthine in the urine. The differential diagnosis of hypouricemia includes many disorders such as uricosuric drugs, a specific defect in uric acid reabsorption from the tubule as reported by Praetorius and Kirk (1) and the Fanconi syndrome such as heavy metal intoxication or Wilson's disease. These are associated with an increase in uric acid excretion however while the association of hypouricemia and hypouricosuria in conjunction with xanthinuria is an expression of xanthine oxidase impairment either primary or induced by enzyme blockers such as allopurinol.

The mode of inheritance of xanthinuria has not been clearly established. It is generally held that enzyme deficiencies are related to a recessively inherited defect in enzyme production. Studies bearing on this point in xanthinuria are somewhat confusing. Consistent with this mode of inheritance, two siblings have been reported on two occasions to have the disease and there was no vertical transmission. Heterozygotes have not been clearly identified although in one family a third sibling had normal uric acid levels and excretion but increased oxypurine excretion. It has been thought that this could be a heterozygote but other heterozygotes have not been identified. Rapado reported a family in whom he felt there was dominant inheritance with low penetrance on the basis of two siblings affected in two successive generations (2).

We have had the opportunity to screen a large family of persons between their 4th and 7th decade who were siblings, nephews and nieces of a patient with primary xanthinuria.

Studies Performed

The initial patient, 63 y.o. P.J. (II_7, Fig 1), was discovered on a routine examination for prostatism to have arthritis in his back, left elbow and shoulder. A serum uric acid was 1.1 mg. percent, while his 24-hour urinary uric acid was only 32 mg/24 hr. A family analysis showed that he had six siblings and 20 nieces and nephews. His mother and father were deceased but his father P.J. had had his right kidney removed at this institution 32 years ago.

An initial screen of the six siblings was performed on an ad libitum diet measuring serum uric acid, 24-hourly excretion of urinary uric acid and oxypurine excretion. Containers with 10 cc phenol as preservative were sent by mail. Two patients had moderately or only minimally elevated oxypurines, but normal serum and urinary uric acid. Therefore, repeat urine samples were obtained from these patients and their children while on a meat free diet for two days, avoiding coffee and tea during the collection period.

All subjects were sent a questionnaire concerning any illness they had with respect to the major bodily systems. Specifically symptoms of arthritis, heart disease, kidney stones, hypertension were sought. When positive, a telephone conversation confirmed the nature of their illness.

Three subjects were given 32 mgm phenobarbital three times daily for five days. Urinary uric acid was monitored before and after drug therapy.

Uric acid was determined by uricase method (3), total oxypurines were determined by the method of Klinenberg (4).

Results

There were two subjects who clearly had primary xanthinuria (Table I), (Fig 1). P.J. and E.S. had 24-hour urinary excretion of uric acid of 36 and 19 mg/24 hr respectively and their serum uric acids were 1.1 and 0.7. After being on three days of a meat free diet their urinary uric acid values were 32 and 6.5 mg/24 hr, and their serum uric acid levels were 0.5 and 0.7 mg percent. In P.J. oxypurine excretion was unchanged by altering dietary urate being 409 and 407 mg/24 hr. In E.S. however oxypurine excretion fell from 263 mg/24 hr to 198 mg on a restricted diet.

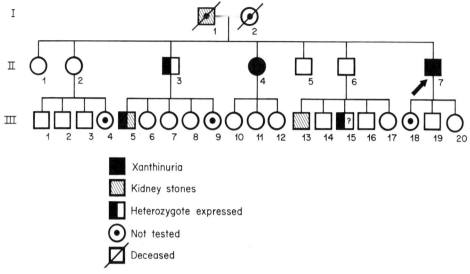

Figure 1.   Kindred with xanthinuria

Of the other five siblings only one (II_1) had absolutely normal values (11.0 and 18.0 mg/24 hr) on both occasions. One patient (II_3) had clearly elevated oxypurine excretion both before (94.2 mg/24 hr) and after (78.0 mg/24 hr) meat and coffee restriction. The other three siblings $II_{2,5,6}$ had one value slightly elevated and one normal value (Table I). Two fell and one rose after dietary restriction. As a total, four of six patients with an abnormal value fell after dietary restriction.

Most kindreds reported in the literature and our data are most consistent with an incompletely recessive mode of inheritance. If that is true the children of $II_{4,7}$ are obligate heterozygotes. They are identified as $III_{10-12;18-20}$ and none of them had any abnormality of urinary oxypurine excretion or uric acid excretion.

Table I shows the probability of being a carrier in the various parts of the kinship assuming a recessive mode of inheritance and assuming that generation I are both heterozygotes. It is clear that there is no general decrease in urine uric acid nor general increase of oxypurine excretion in the obligate or high risk carriers than with the persons of relatively low risk or compared to the upper limit of normals of 20 mg/24 hr (4). Fifteen of the 25 relatives are at risk for a heterozygote state assuming two-thirds of unaffected siblings are at risk and one-half of their off-springs are at risk. The chances of the off-spring of the unaffected siblings being affected is one-third.

We are assuming that one-half of the children of II3 are at risk. Our data show that while on a meat free diet only three of these persons had an elevated oxypurine excretion.

XANTHINURIA

Patients	Urine uric acid, mg/24 hours	Urine oxypurine, mg/24 hours	Serum uric acid, mg/dl
Homozygotes = 2			
II-4	59	198 (263)	0.7
II-7	29	407 (408)	0.5
Mean	44	302	
Obligate heterozygotes = 6			
III-10	485	7.7	
III-11	807	11.1	
III-12	475	1.1	
III-18	379	4.1	
III-19	535	6.1	4.9
III-20	566		4.3
Mean	541	6.0	
Probable heterozygotes 2:3 = 2.6			
II-1	501	11.0 (18.7)	4.6
II-2	513	9.7 (36.8)	4.8
II-3	614	78.0 (94)	5.0
II-5	608	8.7 (22.6)	6.2
II-6	254	4.4 (22.6)	5.3
Mean	376	26.7	
Possible heterozygotes 1:3 = 2.3			
III-1	582	12.3	
III-2	712	15.0	
III-3	713	3.7	
III-4			
III-13	517	1.1	6.4
III-14	713	3.7	
III-15	441	37.8	
III-16	437	6.7	
III-17			
Mean	588	11.5	
Questionable heterozygotes 1:2 = 2			
III-5	470	116 (179)	6.0
III-6	481	8.4	
III-7	362	19.4	
III-8	297	9.2	
III-9			
Mean	403	38.3	

Table I. Uric acid and oxypurine excretion in family of patients with xanthinuria. Brackets ( ) indicate values on normal purine diet (see text)

Table II shows the relatives at risk in the published reports.
There have been 50 patients tested in whom data can be analysed.
As close as can be calculated 36 of these should have been heterozygotes based on an autosomal recessive inheritance. It can be
readily appreciated that in no case was an abnormality of serum or
urinary uric acid perceived. Only one other patient besides our
series was reportedly tested for increased urinary oxypurine and
it was elevated. Tobias states in his review that oxypurine excretion was elevated in the children of two siblings with xanthinuria
from Johannesburg although no data is available (5). Three of the
15 patients at risk in our kindred had elevated oxypurine on a low
meat diet.

The medical problems of these patients have been of some
interest. There have been very few specific complaints although
deposits in tissues have been thought to cause arthritis in at
least one patient. The medical problems of our patients were
rather unremarkable. Of the siblings involved two had hypertension
and P.J. had bursitis of his left shoulder and elbow which cleared
spontaneously. The other siblings had arthritis 3/5, hypertension
1/5, duodenal ulcer, diabetes mellitus, hypothyroidism, lung disease, heart failure each in one instance. Three patients in this
kindred had renal stones. P.J. (I_1) had a "staghorn" calculus and
nephrectomy. He had had a long history of urinary tract infection.

HETEROZYGOTE EXPRESSION

Author	Total no. of relatives	Patients at risk	Decreased Serum UA	Decreased Urine UA	Increased urinary oxypurine
Dent & Philpot, 1954	13	10.3	0	0	
Cifuentes, 1967		1			1/1
Sperling, 1970	3	2	0	0	
Engelman, 1964	1*	1	0	0	
Curnow, 1971	2	2	0	0	
Tobias, 1972	?	?			?/?
Aquazian, 1964	6	5	0	0	
Wilson, 1973	25	15	0/22	0/22	3/21
Total	50	36.3	0	0	4

*Tissue xanthine oxidase assay = 0/1.

Table II. Serum and urine urate and urine oxypurine excretion in
heterozygotes for xanthine oxidase deficiency.

Stone analysis showed only magnesium ammonium phosphate and no xanthine was present. R.J. (III5) had renal colic and a stone by x-ray but no stone was recovered and G.J. (III3) had a calcium oxalate kidney stone.

In an attempt to induce increased enzyme formation to see if there was a low level of enzyme or whether the enzyme was structurally impaired we gave phenobarbital. We were unable to see any increase in the two patients with xanthinuria and also in patient II2 who was a presumed heterozygote based on her original oxypurine excretion of 37 mg/24 hr. This is consistent with a structurally abnormal enzyme but xanthine oxidase is apparently a cytosomic enzyme and may not be induced with phenobarbital.

Discussion

The presumption here is that we are studying a family with primary xanthinuria secondary to a defect in xanthine oxidase. The urinary excretion data of uric acid and oxypurine in conjunction with the extremely low serum urate speak strongly for this. However, we have not measured the xanthine oxidase levels in the tissue of these patients leaving some small measure of doubt as to the true etiology of these deranged values.

The mode of inheritance of xanthinuria is one of the major concerns of this study. The fact that two siblings are involved and there is no vertical transmission make an autosomal recessive inheritance likely. Patient II3 could have been recessive with incomplete penetrance or a heterozygote. His son (III5) is almost certainly a heterozygote with virtually the same expression as his father i.e. normal serum urate (6.0 mg percent), urinary urate (609 mg/24 hr) and elevated oxypurine excretion. This makes it likely that II3 is a heterozygote and is strong evidence that the heterozygote can be identified at least in some cases.

The recognition of heterozygotes in our studies while on a meat free diet was only three out of 15 patients at risk, or 20 percent. This is poor although obviously better than serum or urinary uric acid which is always normal in the heterozygote. Our experience with the siblings of P.J. suggest that oxypurine excretion is higher on a regular diet with four out of five having slightly high values. However, it is not clear what excretion in normals is on a regular diet. It is quite possible that RNA loading or a high meat diet may increase the yield in determining heterozygotes. On the other hand the fact that some heterozygotes show increased oxypurine excretion while others do not may suggest there are two alleles for xanthine oxidase, one governing synthesis and the other enzyme expression or repression.

The clinical expression of any disease associated with this

condition might be expected to be apparent in this kindred who are
in their 60's and 70's for the second generation and 30's and 40's
for the third generation. Of the cases previously reported three
had xanthine stones. Neither of our patients had stones. There
were three stones in the kindred two of whom were clearly heterozygotes. It is unlikely that any of these stones contained xanthine. Two were analysed and the third was radiopaque. Increased
uric acid excretion appears to be associated with increased
calcium oxalate stones, but that association has not been reported
to date for increased xanthine excretion, either in these patients
or those on allopurinol.

Three patients have been reported with a myopathy and at least
one with polyarthritis. We did see arthritis in 3/7 siblings and
III_5 all of whom had increased oxypurine excretion. There was no
evidence of a myopathy with the possible exception of II_2 who had
heart failure, but she also had hypertension.

References

1.   Praetorius E. and Kirk, J.E.: Hypouricemia with evidence of
      tubular elimination of uric acid. J.Lab Clin Med 35:865-68,
      1950.
2.   Rapado, A., Castro-Mendoza, H., Cifuentes, D.L.: Renal stone
      research symposium. Absts. p 29 Madrid, Spain Sept. 1972.
3.   Liddle, L., Seegmiller, J.E., Laster, L.: The enzymatic
      spectrophotometric method for determination of uric acid.
      J Lab & Clin Med 54:903-13, 1959.
4.   Klinenberg, J.R., Goldfinger, S., Bradley, K.H., Seegmiller,
      J.E.: An enzymatic spectrophotometric method for determina-
      tion of xanthine and hypoxanthine. Clin Chemistry 13:834-46,
      1967.
5.   Tobias, P.V.: Some genetical aspects of hyperuricaemia and
      xanthinuria. S African Med J 46:552-4, 1972.
6.   Dent, C.E. and Philpot, G.R.: Xanthinuria. An inborn error
      of metabolism. Lancet 1:182-85, 1954.
7.   Cifuentes, D.L. and Castro-Mendoza, H.: Xanthinuria familiar.
      Rev Clin Esp 107:244, 1964.
8.   Sperling, O., Liberman, U.A., DeVries, A., Frank, M.: Xan-
      thinuria: An additional case with demonstration of xanthine
      oxidase deficiency. A.J.C.P. 55:351-54, 1971.
9.   Engelman, K., Watts, W.E., Klinenberg, J.R., Seegmiller, J.E.,
      Sjordsma, A.: Clinical, physiological and biochemical studies
      of a patient with xanthinuria and pheochromocytoma. Am J
      Med 37:839-61, 1964.
10.   Curnow, D.H., Masarel, J.R., Cullen, K.J., McCall, M.G.:
      Xanthinuria discovered in population screening. BMJ I:403:71
11.   Ayvazian, J.H.: Xanthinuria and hemochromatosis. NEJM 270,
      18, 1964.

Glycogen Storage Disease

AN UNUSUAL CASE OF GLYCOGEN STORAGE DISEASE

S.W. MOSES, M.D.

Department of Pediatrics B and Pediatric
Research Laboratory, Soroka Medical Center
and the University of the Negev.*

Glycogen Storage Disease Type I (G.S.D.I) is
characterised by a block in the final common pathway
of glucose liberation from the liver leading to
abnormal deposition of liver glycogen.   The pertinent
metabolic features include fasting hypoglycemia, lactic
acidemia and increased free fatty acids.   A severe
tendency towards acidosis, hyperuricemia, hypertriglyceridemia with an abnormal pre- β band are
frequently observed.   Glucagon administration does not
provoke hyperglycemic response, but causes a further
rise in blood lactic acid.   No hyperglycemic response
can be elicited after galactose, fructose or glycerol
administration.

Characteristically the disease is associated with
liver glucose-6-Phosphatase deficiency.   However,
cases are on record who show typical clinical and
biochemical features of G.S.D.I yet, normal glucose-
6-phosphatase has been found by in vitro measurements
of liver enzyme.

The following presentation shows that, by applying
more sophisticated metabolic parameters, the former
interpretation of the pathogenetic mechanism operating
in such cases may not apply and other explanations have
to be sought for.

Y.B. was born in 1965 as the product of a first
cousin marriage between two Israelis from Iraqui

*Supported in part by a grant from the Israel Council
for Research and Development.

extraction who have two older children.   Both parents
and siblings are healthy.

The patients illness first became manifest at the
age of three months when he suffered from fasting
hypoglycemia with convulsive attacks which were
preventable only by two hour feedings of carbohydrates.
In addition to profound hypoglycemia elevated levels
of lactate and uric acid were found.   Epinephrine or
glucagon administration did not result in a rise of
blood sugar.   A large liver could be palpated, which
on histological examination of a needle biopsy revealed
to be glycogen and lipid laden.

Leucocyte enzymes revealed normal phosphorylase,
amylo-1,6-glucosidase and a somewhat low α glucosidase
activity.   Red blood cell glycogen was normal.
(Table 1).

A repeat glucagon stimulation test performed in
another hospital at the age of 18 months resulted in
no rise in blood glucose, although there was evidence
of phosphorylase activation with glycolysis in the
form of a rise in blood lactate from 20 to 80 mg%
during the procedure.   Both galactose and glycerol
which are normally converted by different pathways to
glucose-6-phosphate and then to free glucose via the
glucose-6-phosphatase enzyme, failed to reverse the
rapid post prandial decline in blood glucose. (Table 2).

The pertinent biochemical data of this case are
presented in Table 3.   The high levels of uric acid
found in this case responded to the administration of
allopurinol.

Studies on insulin and growth hormone levels
following an i.v. glucose tolerance test (G.T.T.)
Table 4, shows that blood glucose elicits a marked
response in serum insulin, unlike responses found in
classical cases of G.S.D.I as described by Lockwood.
Similarly, the G.T.T. did not show a diabetic pattern
as seen in insulinopenic G.S.D.I patients.

In order to prolong the fasting tolerance and
alleviate the hypoglycemic attacks, diazoxide, an agent
which lowers insulin production and may also reduce
peripheral glucose utilisation, was initiated.   Metabolic studies during the administration with this drug,
presented in Fig. 1, indicate that fasting tolerance

T A B L E   I.

LEUCOCYTE   ENZYMES

	PATIENT	NORMAL CONTROL
AMYLO-1.6-GLUCOSIDASE $\%/10^{10}$ CELLS/HOUR	8	16
PHOSPHORYLASE /uM Pi/10^{10}/min	57	51
α GLUCOSIDASE /uM MALTOSE/10^{10}/min	0.62	2.95
RED BLOOD CELL GLYCOGEN /gHb	122	

T A B L E   2.

TEST	RESULT
EPINEPHRINE)) GLUCAGON)) GALACTOSE)) GLYCEROL)	NO RISE IN BLOOD GLUCOSE

T A B L E   3.

BLOOD   CHEMISTRY

(3 HOURS POST PRANDIAL)

GLUCOSE	0-35 mg%
LACTATE	10-43 mg%
URIC ACID	10.4 - 11.9 mg%
URIC ACID AFTER ALLOPURINOL	6.0 - 8.2 mg%
CHOLESTEROL	258 - 311 mg%
TOTAL LIPIDS	1230 - 2870 mg%
SGOT	43 - 115 UNITS
SGPT	61 - 83 "

T A B L E   4.

i.v. G.T.T. 0.5 g/Kg AFTER 2½ HOUR FAST

TIME in min.	GLUCOSE mg%	INSULIN /uV/ml	GROWTH HORMONE m /ug/ml
0	62	52	2.7
3	171	39	3.0
6	212	22	1.1
10	167	50	0
15	171	10	1.4
30	120	10	1.5
45	93	10	1.2

was somewhat improved and no lactate rise was demonstrable as expected.   A characteristic rise in F.F.A. as blood glucose dropped, is noticeable.

This case qualifies by clinical and biochemical criteria as G.S.D.I.   Accordingly it was a considerable surprise to find in a liver biopsy that both by histo-chemical and standard biochemical procedures, normal activity of glucose-6-phosphatase was demonstrable. The liver tissue contained 8% glycogen.   There was considerable fat accumulation but no evidence of cirrhosis.   Similarly normal activities of liver phosphorylase and amylo-1,6-glucosidase were found.

It has been suggested by Senior to classify such cases as type Ib with the assumption that although glucose-6-phosphatase is present in tissue it is not available for metabolism because of local factors. (2)

J. Schrub et al found abnormal vesiculation in the rough endoplasmic reticulum of the liver in a similar case (1).

L. Hue and H.G. Hers recently described a method employing double labelled glucose (2 - ^{3}H and U-C^{14}) to assay for the activity of liver glucose-6-phosphatase in vivo.   The test is based on the principle that a differential in the label of ^{3}H as compared to C^{14} glucose is produced during the recycling of glucose at the level of the phosphohexose-isomerase reaction.

As a result of this recycling ^{3}H is replaced by non-radioactive hydrogen, consequently the ^{3}H/C^{14} ratio of glucose liberated from the liver into the circulation decreases progressively.   Obviously this decrease of ^{3}H/C^{14} ratio in blood glucose requires the operation of a functional glucose-6-phosphatase.

Therefore, in the absence of glucose-6-phosphatase as found in G.S.D.I it is expected that this decrease will not be present since double-labelled glucose present in the circulation will not be diluted with glucose liberated from the liver after having lost ^{3}H while passing through the isomerase reaction.

It was, therefore, considered to be of considerable interest to perform this test on our patient.   If indeed he represents, as formerly suggested, a case in

T A B L E   5.

$^{3}H/^{14}C$ RATIO IN BLOOD GLUCOSE

(% of initial value)

	0 min	30 min	60 min
CONTROL (n=8)	100	78 ± 4	60 ± 7
TYPE I (n=5)	100	101 ± 2	97 ± 3
B.Y.	100	13	too low to be detected

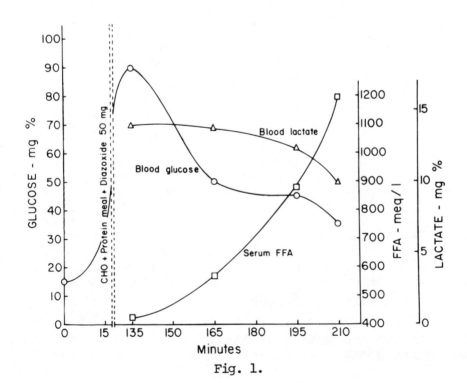

Fig. 1.

whom glucose-6-phosphatase is not functional in vivo, the results of this test are expected to be similar to the one found in G.S.D.I.

The performance of this test was made possible by the kind help of Dr. Hue from the biochemical laboratory of Dr. Hers in Louvain, Belgium, and the results are shown in Table 5.

The most surprising rapid decrease of the ratio suggests both that a most active recycling occurs and that recycled intermediates can be liberated as glucose into the circulation. This is most confusing as all the other clinical and biochemical evidence accumulated in this case, point to a functional defect in the capacity of this patient to dephosphorylise glucose.

It is unlikely that an increased brancher-debrancher shuttle which results in the liberation of small amounts of glucose without passing glucose-6-phosphatase could account for these figures. Similarly to assume that an increased activity of glucosidase can be incriminated to explain these findings is not likely. Furthermore, low values of leucocyte α-glucosidase have been documented in this case. It is, therefore, difficult on the basis of our present knowledge of existing enzymatic defects in the pathway of carbohydrate metabolism to explain this puzzling case.

It may be pertinent to quote in this respect William Harvey who stated about 350 years ago that nature hides some of its mysterious secrets far away from the trodden path. This may be one of them.

R E F E R E N C E S

1.   J. Schrub et al: Ped. Research 50; 66, 1973.

2.   B. Senior and L. Loridan: Studies of Liver
       Glycogenosis with particular reference to
       the metabolism of intravenously administered
       glucose. New Engl Jnl Med 958; 279, 1968.

HYPERURICEMIA AND DICARBOXYLICACIDURIA IN GLYCOGEN STORAGE DISEASE

J.A. Dosman, J.C. Crawhall,G.A. Klassen and O.
A. Mamer
Divisions of Cardiology and Clinical Biochemis-
try, McGill University Clinic, Royal Victoria
Hospital, Montreal, Quebec, CANADA

An 8 year old girl with clinical and laboratory findings compatible with Glycogen Storage Disease Type I was admitted to the Investigative Unit for study of the origin of her hyperuricemia. Outstanding physical signs on this child were her protruberant abdomen with greatly enlarged liver. Her weight was at the 9th percentile for her age and her height at the 3rd percentile. Laboratory findings showed a serum cholesterol of 390mg% and trigly-cerides of 1630mg%. The total fasting free fatty acids were 1.16μM/ml and the blood lactate on repeated deter-minations ranged from 5 to 7.6μM/ml. Serum uric acid le-vels ranged from 6.5 to 7mg% (Uricase method). Hypogly-cemia was not an important feature of her disease at this stage. While in the hospital she was maintained on a low purine calorically adequate diet but this was supplemented by various carbonated beverages and candy bars. Her diet was estimated to contain 45gm of protein, 50mg of purine, 70gm of fat and 220gm of carbohydrate. Uric acid-2-C^{14} was injected and complete urine collections made at 12 hourly intervals for the first 3 days and then subsequent-ly at 24 hourly intervals for the next 4 days. The uri-ne was collected in glass jars containing a solution of lithium carbonate and uric acid was subsequently isolated and purified by conventional procedures. The total radio-activity of the urinary uric acid decreased in a simple exponential manner, and the appropriate calculations showed that the uric acid pool size was 821mg or 36mg per kilogramme body weight and the turnover of uric acid was 940mg/24hours. Studies in normal subjects were car-

ried out as part of a different project but these were
in adults and not strictly comparable to the findings in
this particular patient. However, the adult values that
were obtained were compatible with other published data
for uric acid turnover in normal adults and the values
found in this child confirmed the previous observation
of Kelly et al. (1968) that there is an excessive production of uric acid in Type I Glycogen Storage Disease.
This patient also had high circulating lactate levels but
the level of ketone bodies was very minimal, a finding
compatible with that of Fernandes and Pikaar (1972).

During the course of this study it was decided to
investigate the excretion of urinary organic acids by
the technique of GC-Mass Spectrometry. Large quantities
of lactic acid and beta-hydroxybutyric acid were found
in the urine and also some alpha-hydroxybutyric acid.
In the region of the chromatogram corresponding to the
larger molecular weight organic acids, we were surprised
to see unusually large quantities of glutaric, adipic,
pimelic, suberic, azelaic, and sebacic acids (Fig.1).
Many analyses of this type were carried out which showed
that the quantity of aliphatic dicarboxylic acids being
excreted by the child was quite variable. The excretion
of adipic acid appeared to be a fairly non-specific phe-
nomena but the excretion of dicarboxylic acids of chain
length C_7, C_8, C_9 and C_{10} was characteristically different from other patients we had examined with other types of metabolic acidosis or ketosis (Table I). Further
studies were somewhat delayed by technological difficulties. One of these was that ordinary gas chromatographic techniques are somewhat unreliable for the identification of these acids by their elution time and hence
positive identification required the use of the GC-Mass
Spectrometer. There are certain difficulties in quan-
titating the values obtained by this instrument even
with the use of an internal standard. When these problems had been solved we went back to re-examine the dicarboxylic acid excretion of this child and also the urine of the child's older brother who also had the disease and another non-related patient. One other patient
diagnosed as having Type III Glycogen Storage disease
was also studied. Although all these children showed
abnormal excretion of urinary dicarboxylic acids at the
time of their original investigation, the excretion of
these acids on a semi-quantitative basis had obviously
decreased one year later and accurate quantitation after
appropriate calibration of the instrument using internal
standards showed that the quantity of dicarboxylic acids

TABLE 1

QUANTITY OF URINARY DICARBOXYLIC ACIDS EXCRETED EXPRESSED AS A PERCEN-
TAGE OF TOTAL PEAK HEIGHTS ELUTING FROM THE GAS-CHROMATOGRAM AFTER SUCCINATE

	Sex	ADIPIC	SUBERIC	AZELAIC	SEBACIC	TOTAL
GSD-I						
P.M.	M	2.8	7.2	16.7	3.3	30.0
E.M.	F	6.5	15.1	10.6	–	32.2
D.B.	F	4.7	15.0	21.8	11.4	52.9
GSD-III						
S.C.	M	8.0	6.8	–	–	14.8

This percentage was obtained by adding up the peak heights of all the urinary dicarboxylic acids on the gas chromatogram and dividing by the total of
all the organic acid peaks eluting after succinic acid and expressing the result as a percentage.

The same analysis and calculation was carried out on urine from 7 normal
control children of comparable age which gave values for total dicarboxylic
acids of 14, 0.4, 0, 0, 1.2, 4.4, 5.5.

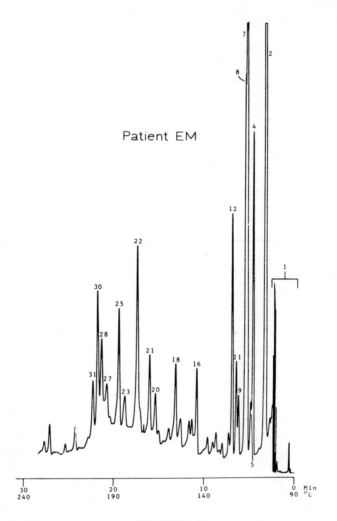

Patient EM

FIGURE 1

Total ion current record of peaks eluting from the
gas chromatographic inlet in the silylated mixture of
organic acids extracted from the urine of patient E.M.
The peaks are identified as the following acids by comparison of their retention times and mass spectra with
those of authentic compounds: 1, solvents and silylating
reagents, separator valve attenuated; 2, lactic; 3, α-
hydroxyisobutyric; 4, α-hydroxybutyric; 5, α-hydroxy-α-
methylbutyric; 6, α-methyl-β-hydroxybutyric; 7, β-hydro-
xybutyric; 8, α-hydroxyisovaleric; 9, α-methyl-β-hydro-

xybutyric; 10, acetoacetic, early isomer; 11, β-hydroxy-
isovaleric; 12, ω hydroxypentanoic, unknown isomer; 13,
acetoacetic, late isomer; 14, succinic; 15, fumaric; 16,
glutaric; 17, β-methylglutaric; 18, adipic; 19, unknown
methyl-substituted adipic isomer; 20, pimelic co-eluting
with m-hydroxyphenylacetic, ratio ranging 2:1 to 4:1 approximately; 21, ρ-hydroxyphenylacetic; 22, suberic; 23,
homovanillic and vanillic co-eluting in varying ratios;
24, (dihydroxyphenyl) ethanol, unknown isomer; 25, azelaic; 26, hippuric; 27, m-hydroxyphenylhydracrylic; 28,
sebacic; 29, unknown, no sebacic co-eluting; 30, ρ-hy-
droxyphenyllactic; 31, 3-indolylacetic; 32, 5-hydroxy-
indolylacetic.

TABLE 2

EXCRETION OF ADIPIC AND SUBERIC ACIDS IN URINE OF

PATIENTS WITH GLYCOGEN STORAGE DISEASE AND NORMAL

CHILDREN OF SIMILAR AGE

(mg/gm creatinine)

GSD-I	ADIPIC	SUBERIC	TOTAL
E.M.	24	11	35
D.B.	27	8	35
GSD-III			
S.C.	12	10	22
NORMALS			
1.	8	0	8
2.	1	2	3
3.	3	0	3
4.	18	3	21
5.	0	0	0
6.	0	0	0

being excreted by two of the Type I patients was 35mg/g
of creatinine and in the Type III patient, 22mg/g of creatinine (Table 2).   These values were 3 or 4 times higher
than 5 of our normal control patients although there was
one normal control patient who appeared to have a persis-
tant elevation of urinary adipic acid.   All these patients
were being cared for by the same pediatrician and we believed that a possible explanation for their originally
more prominant dicarboxylic acid excretion which subsequently declined was that the physician changed their
diets from high to low carbohydrate intake in an attempt
to control their high triglyceridemia.   Previous studies
by Verkade et al. (1948), showed that normal subjects excreted large amounts of urinary dicarboxylic acids when
fed exogenous triglyceride plus carbohydrate but that
they excreted little dicarboxylic acids when triglycerides were fed during carbohydrate deprivation.

Several other investigators have demonstrated that
omega-oxidation of fatty acids can occur giving rise to
urinary dicarboxylic acids.   Hypertriglyceridemia per se
does not seem to be related to the excretion of these
acids in the adult.   However, in the endogenous carbohydrate deprivation of Glycogen Storage Disease the attempts
of the body to maintain glucose homeostasis by gluconeo-
genesis may induce a situation whereby omega-oxidation
of partially degraded long chain fatty acids occurs.

Kelley, W.N., Rosenbloom, F.M., Seegmiller, J.E.
and Howell, R.R.; J. Pediat. 72, 488, 1968.
Fernandes,J. and Pikaar, N.E.;   Arch. Dis. Child.
47, 41, 1972.
Verkade, P.E., Vander Lee, J., and Elzas, M.; Biochim. Biophys. Act., 2, 38, 1948.

We acknowledge the continuous cooperation of Dr.
P. Neumann who was responsible for the children's care
and the Canadian M.R.C. for financial support (Grants
No. MA-3331, MA-3719).

SUBJECT INDEX

Pages 1 through 366 will be found in Volume 41A, pages 367 through 833 in Volume 41B

Adenine
  HGPRT deficiency
    effect of dietary purines in,204
    on enzyme thermostability,206
    on phosphoribosylamine synthesis,248-269
  Lesch-Nyhan syndrome(LNS)
    therapy with,669-679
    acute renal failure, 671-675
    behavioral changes,677-679
  nucleotides, renewal in erythrocytes from liver adenosine,129-139
  platelets nucleotides, incorporation into,159-162
  PRPP, effect on intracellular level,95
  PRT, destabilization to heat,215-220
  xanthine dehydrogenase inactivation,68
Adenine phosphoribosyltransferase (APRT)
  alternate aggregation states,29
  benign symmetric lipomatosis,514
  deficiency,23,319-326, 327-331,333-339
  properties of mutant enzyme,321-322
  inactivation by freezing and thawing and stabilization by PRPP, 15-18
  Lesch-Nyhan syndrome
    elevated levels of activity,23
    fibroblasts,17
    resistance to destabilization to heat by hypoxanthine, 215-220
  properties in hemolysates,215-220
  partial HGPRT deficiency
    properties in hemolysates,212
  platelets,159-162
  purification,subunit structure and substrate specificity,23-32
Adenosine
  erythrocyte,incorporation into nucleotide pools,122
  liver,role in renewal of erythrocyte nucleotides,129-139